# The Emergent Mind

Evolution-The Wisdom of Nature

Frances Grace Rogers

Copyright © 2010 Frances Grace Rogers
All rights reserved.

ISBN: 1451569653
ISBN-13: 9781451569650

# Acknowledgments

I wish to express my gratitude to Jerusha Golden for reading my work, especially in its initial stages. Her feedback was valuable in helping me to clarify my words and thoughts. I am grateful, as well, to Ann Horne for generously offering her time to read and provide her impressions. The encouragement of Jerusha, Ann, and Ed, who was frequently my sounding board, supported me to bring my ideas to fruition.

For Carla, Miles, Emily
and
For Charlie

One touch of nature makes
the whole world akin.

William Shakespeare

Recorded musical compositions by the author are available at URL:
http//www.soundcloud.com/the-emergent-mind

# Preface by the Author

The picture on the cover of *The Emergent Mind* is one of my paintings and, like my poetry, spontaneous writings, and musical compositions, art played a major role in my evolution out of the Flatland of closed systems. The coming of the creative is my story of emergence.

Sometimes what is missing in a story is of major significance. Conspicuous to mine, I have not used the word *journey* that is frequently employed in memoirs; it implies movement from one measurable place to another as does the word *progress*—the linearity which is the cultural bias throughout most of the world. It is progress that has not taken us very far in humanitarian terms and instead, has left its destructive mark on society and our planet. What has been damaged by humans during that *progress* must be corrected by humans.

My story is one of an unfolding that refutes that cultural bias. And *progress*, in this writing, is reserved for the process of maturation—an upward development.

For most of my life, I believed in and trusted the basic goodness of people, so a major part of my "memoir" revelations became the facing of the truth of evil and how it affected the multiple dimensions of my life.

*Evil* is not a word I use lightly and certainly not in religious terms, but rather in a concept defined by the words of Albert Schweitzer regarding civilization: "Reverence for Life affords me my fundamental principle of morality, namely, that good consists in maintaining, assisting and enhancing life, and to destroy, to harm or to hinder life is evil."

Evil—the product of closed systems invented by humans—is threatening the survival of humanity, and a major underlying factor is complacency. Edmund Burke's wise words reflect my sentiment: "All that is necessary for the triumph of evil is for good men [and women] to do nothing."

I still believe in the basic goodness of people. I also believe in the human potential to curb those destructive forces by mending the splits and divisions both within and among us, awakening the possibilities of the Emergent Mind.

I

# Golden Frogs and Such

The following passage was originally titled *My Golden Frog*, and I wrote it in a matter of a few hours on April 14, 1993. The words came as though dictated from an unknown source:

*It was in my dark days, hours, minutes with no name, time stilled except for habits calling me to meet mechanical needs; eating, excreting, walking on two legs; habits, like breathing, to keep my body alive and to reassure curious, concerned eyes watching me. I was elsewhere, returning to my robot program to mouth those meaningless words of conscious conversations, words to appease, words formed to prevent shaking the intricate illusion of sanity or startling those who could not comprehend the screams from my soul.*

*In the darkness I was not alone, for all the ghosts of the past rushed in waves, those chaotic images of memory that seemed more real than reality. I would sink further into the depths, and my senses, startlingly alive, twisted my tongue to form the hollow sound, "My soul is black!"*

*So it was black with darkness, and menacing, like all souls that have been shut to the light of consciousness to maintain the myths humans have concocted and live.*

*My soul's voice dictated poems, and I counted syllables to match the rhythmic pulses of my heart, verses predicting all that was ahead of me. They were poems that laid to waste all past stories, creating a new one by which I am to live and tell.*

*In the light of consciousness, I attempted to accommodate the old mechanistic story, only to be driven back into the darkness by a correcting voice, words that kept me humble and would not allow me to stray from my task. Following the call to higher order, releasing my spirit from the clutches of obsession, I was to pay the price of freedom, the freedom "to be," as the spirit of humans is meant to be.*

*In those days of timelessness, when all senses learned were lost, my golden frog appeared, a sign that my conscious, mechanical eye suspected as madness. It was poised silently on my doorstep and in the twilight seemed only a shadow but for the glow of golden stripes down its back. I knelt, and my two eyes, seemingly open to separate worlds, argued over the metallic, luminous, olive green skin dotted with gold and shimmering golden orbs above its eyes. Such color, such beauty, a decorator would desire to mimic.*

*To know the beauty of a frog, of being alive but stilled by its wounds, for me to capture proof or to let go, seemed a miracle of the darkness. No human had heard the longing of my soul, my unsung prayer that rose to the ether for 40 years for reassurance of existence. No human could promise the gift of illumination, only nature perceived by my soul's eye, a calming force that stilled my fear.*

*Golden Leopard Frogs exist in the southeast region of the United States yet are rarely seen because conditioned eyes tell the story of ugly frogs and warts and look for magic to exchange them into princes, titles, power, and coins.*

*The darkness passed, time returned, and I perceived a different world. I reeled from the sight, my brain spinning to absorb the light of unified consciousness; no secrets left to haunt me. It was not the beauty of golden frogs or the rapture of mystic, twinkling obsession that greeted me. It was the remnants of human waste piled in chaotic heaps, reeking with the stench of carcasses rotting in the sun. It was my debris and that of generations before me that had been kept alive like poison mold. I mourned each fragment of bone and flesh and blood extricated from those piles. Piece by piece they were buried in graves marked with stones smoothed by my river of tears.*

*Rage cried too, for at its root was a deep dark well of fuel, a thousand liquid thoughts and feelings that surfaced to burn in scorching flames. So let it be; fire and water cleansed me. Time, sweet time, allowed my wounds to heal.*

*That golden frog is alive in memory now, those points of light that do, with selective magic, gather seeds to sprout in ashes and graveyards; that golden frog which in the darkness did shine and by its presence upheld the other signs given me, the signs the Spirit gives when humans are open to receive them.*

*There is order in revelation, a precise order in what the timely eye calls madness, an order which bears a weight of knowledge and even heavier, the burden of the telling in linear terms of dictionaries. It is a burden lightened, though, by human curiosity about other worlds which instinct, the prophetic self, gobbles to feed the starving soul.*

That unknown source of dictation is no longer a mystery, for I now know and accept the creative process which is oftentimes stranger than fiction.

The content of *My Golden Frog* is a vivid description of multiple layers of time, encompassing elements from where I was in 1993, where I had been previously, and where I would wind up on April 14, 2008, the day I began this writing, to further elaborate those revelations. It is a story that could not have been told earlier, for much of that writing was about events yet to be lived and experienced; continuing a course that began over half a century ago in Linden, a mere dot on the map in central Florida.

I recall, as a youth, sitting outside my family's rural home on a grassy knoll and gazing at the vast, clear, evening sky, seemingly alive with the light of the stars and moon. I wondered about my place in all that mystery. In the stillness of the night, a deep sense of yearning seemed to wash over me—to be all that I could be. It was a yearning that culminated in the events inspiring the writing of *My Golden Frog*, what I had been working for, searching for, waiting for, those forty years.

Many times I have attempted to write a depersonalized treatise, to forego the heavy burden of revealing my own story, but finally chose to write in first person—"I." That choice was made when I acknowledged the validity of Friedrich Nietzsche's statement: "Man can stretch himself as he may with his knowledge and appear to himself as objective as he may; in the last analysis he gives nothing but his own biography."[1]

His discerning words are enlightening in many dimensions, including the field of psychology of which I was a part. No doubt Carl

Jung had Nietzsche in mind when he said: "Philosophical criticism has helped me to see that every psychology—my own included—has the character of a subjective confession . . . Even when I am dealing with empirical data I am necessarily speaking about myself."[2]

What I offer in this writing is the process of my movement to wholeness and what I have discovered about oppressive systems men have created that darken the soul. I have also realized the significance of language in maintaining those systems, and paradoxically, the liberating role language can play in bringing light and freedom to the victims.

I have been affected by the words of others in dreadful ways and in profound ways. During that mute period of yearning in my youth "to be," I read a poem by Henry Wadsworth Longfellow, *A Psalm of Life*.[3] It is one of the few things I ever memorized, for verbal memory retrieval is difficult for me. It is an anomaly of my brain that recalls in images. Yet Longfellow's poem was different. I have carried those words with me throughout my life, and frequently, in times of discouragement, have repeated verses one, two, and nine like a mantra:

> Tell me not in mournful numbers,
> Life is but an empty dream!
> For the soul is dead that slumbers,
> And things are not what they seem.
>
> Life is real! Life is earnest!
> And the grave is not its goal;
> Dust thou are, to dust thou returnest,
> Was not spoken of the soul.
>
> Let us then be up and doing,
> With a heart for any fate;
> Still achieving, still pursuing,
> Learn to labor and to wait.

I had no clue how deeply my mantra, and even the omitted verses, were carved into my mind, for they resonated with my inner being.

When my poetry began to flow in 1990 (more dictation-like experiences), the rhythm was strangely familiar, not unlike Longfellow's psalm, the lyrics for which I have since composed the music.

My own poetry simply emerged like *My Golden Frog*—fully formed. It is the creative potential that exists in each of us, a capacity that can lead us to higher levels of consciousness.

As stated in *My Golden Frog*, there is a "precise order" in revelations that evolve consciousness. Joseph Campbell, renowned mythologist and prolific writer, also discovered that order in the myths purported to cross the boundaries of cultures—an archetype for change. Campbell says: "Where we thought to travel outward, we shall come to the center of our own existence. And where we thought to be alone, we shall be with all the world."[4]

Perhaps, in a future generation, it is an archetype that will be outdated, when self-knowledge becomes a way of life. That change will come with open systems and nurturing the wholeness of our young. Archetypes are not born in the psyche but produced by the systems invented by humans.

The hue and cry of millions of people is for change, a sense that we, as a country, are headed in the wrong direction. I join in with my own hope for change—a mending of the splits within us and disunity among us. My hope, though, is not only for our country but for a safer and more egalitarian world where people are mindful of our life-sustaining earth and in caring for each other.

I have come to disagree with a few lines in Longfellow's poem, and one of them is in verse six: "Let the dead past bury its dead." We cannot realize change by simply closing the doors to the past and willing a different future. Rhetoric advocating such idealistic change is mere hype, the product of illusions and false constructs. Before we can achieve true and lasting change, we must not only wake up to where we are, but also explore where we have been to discover the source of deception and insanity of the world in which we live.

## II

## Finding My Metaphor In Flatland

For many years I had no context in which to place my "dark days" described in *My Golden Frog*. During that period it seemed to be me who was insane. In the field of psychology I found one bit of information with some similarity. It is a process Carl Jung termed *the coming of the unconscious*,[5] and the only advice given by Jung was to face the oncoming darkness without prejudice. Plato's famous *Allegory of the Cave*[6] was much more appropriate. It presented images of the actual process of movement out of our illusionary world to the light. (My personal rendition of that parable is recorded at Appendix I.) I had many experiences, though, for which Plato's story and Carl Jung's theory proved inadequate.

It took time and distance before recognizing the metaphor that did fit closely with my own experiences. It has become a priceless gift, for it has served to validate a true awakening on many levels. It is contained in the brief humoresque satire titled *Flatland, A Romance of Many Dimensions*,[7] purportedly the first science fiction novel. It was written by Edwin A. Abbott, master teacher and theologian in 19th century England. It has now regained popularity with new editions read primarily by mathematicians and computer scientists.

It is crucial to address a problem encountered in reviewing Abbot's work. It is a problem inherent in the use of language. It is the process of interpretation or translation, when one sees through the eyes of one's own subjective world.

The initial copy I read is not the same as subsequent publications. The newer editions have been altered slightly, changing a word here and there, which tampers Abbot's original concepts. Even in the 1952 edition I first read, some of Abbot's most salient ideas were omitted. Also, my opinion of Abbott is much more kindly than that of Isaac

Asimov, whose statements in the Forward to the 1983 edition accuses Abbott of sexism. Perhaps that difference is because we were reading different revised editions. I was also left with the impression that Asimov knew nothing of the author and the "biography" Abbot was offering.

Thomas Banchoff, professor at Brown University, in his book, *Beyond the Third Dimension, Geometry, Computer Graphics, and Higher Dimensions*,[8] writes that Abbot led a movement to provide educational opportunities for young men AND women of all social classes. *Flatland*, Banchoff relates, was a result of Abbot's frustration with prevailing social attitudes in 19th century England as well as established views in education and religion. Banchoff describes "this little masterpiece" as one that still speaks clearly to our own day, and it does in ways, in all probability, Asimov or Banchoff did not discern.

I share Abbot's frustration with established views in education, religion, and with prevailing oppressive social attitudes that continue. It is comforting to know Abbot was awake, even a century ago, to the fact that women, who constitute over half of the general population, are oppressed. Injustice, though, is not limited to women but permeates our culture just as it did in Abbot's day. Oppression in any form is a serious and valid problem; therefore my interest is in the process of change, which Abbot prescribes. It is a prescription we do, indeed, need in the present.

While Abbot's satire can be viewed from many perspectives, the "many dimensions" in the title fits faithfully with the dimensionality of his words. His statement that "Flatland has thickness" denotes another problem with linear language, for his metaphor, like any other parable, has multiple layers of meaning. That thickness is often overlooked when one cannot see the forest for the trees.

*Flatland* has more "thickness" than even I recognized upon my initial reading in 1993, for I had not fully understood the phenomenon that art frequently foreshadows discoveries in science. Leonard Shlain, in his book, *Art & Physics, Parallel Visions in Space, Time & Light*,[9] documents many such incidences. Abbot's work not only foreshadowed the discoveries of dimensional geometry, but discoveries of the functioning of the human brain and its role in self-discovery. At the time of the writing of *My Golden Frog*, I did not realize its content was foreshadowing my own self-discovery. Both suggest that the creative process is actually a discovery of creation. It is a very different concept from the corrupted use of the word *creative* which can be assigned to schemes practiced by individuals and on up to corporate or governmental levels.

"A Square," the name Abbot chose for his narrator, begins his tale by projecting the setting forward more than a century in time, to the new millennium—the last day of the year 1999, precisely the year and time I happened to write the following chronicle and put it away. A briefer synopsis would not do justice to "this little masterpiece," for the details are vital in comprehending Abbot's social commentary and movement toward enlightenment. I confess, up front, my focus is on those dimensions reflective of my subjective world.

༄

*Flatland* is divided into two parts: *This World* and *Other Worlds*.

In Part One, the narrator, A Square, initiates the reader into "This World" and the "Flatlanders" whose concept of space is confined to north and south, east and west. Abbott uses geometric figures as representations of the inhabitants; the number and size of the inhabitants' angles strictly divide each caste of society. Circles are the most esteemed of castes, the "high priests" of the various subjects in Flatland. They have so many angles they diffuse into a curved line.

Women, being only straight lines, are the lowest and most pitiable caste, having no angles or any hope of bettering their positions. They are disparaged, abused, and oppressed, forced to announce themselves with the "oh peace" (appease) cry. Their seductiveness is painstakingly described. Their miseries and humiliation are believed to be a result of natural laws or the product of evolution. Their eyes and mouths are perceived as identical, and they evidence no memory, according to A Square, which seems to be a wise "pre-arrangement" in order for them to cope with their miserable state.

The ambient conditions in Flatland, the world of two dimensions (really three, according to the author), is primary darkness, exacerbated by "the fog," which A Square describes as comforting and necessary to the maintenance of the caste structure of society. The "fog" diminishes sensate existence. In the lower classes, feeling is the only means of recognition. A Square, of the professional or gentleman class, acquired, through his education, a keen sense of hearing and sight-recognition to determine the illuminated shapes of other inhabitants of This World. Were it not for the fog, all would appear equal.

A Square recounts a brief time in history when decorative color was introduced, the art that equalized all classes. Art created such havoc that the repressive circles forbade and punished the use of color in the lower castes and reserved it for themselves.

In Flatland it is the circles who dictate conformity: "irregulars," those with uneven sides and shapes are gotten rid of or imprisoned or assigned to some menial, meaningless task. Anyone who entertains the notion of dimensions higher than the two-dimensions of This World is also eliminated or imprisoned.

After familiarizing the reader to life in This World, A Square begins, in Part Two, to narrate his introduction to "Other Worlds," dimensions that lie beyond the limited environment of Flatland. In his movement toward enlightenment, A Square encounters the revealing world of dreams, ideas, along with all the vicissitudes that accompany contradictions to his familiar existence.

In his first dream of dimensions other than his own, A Square is introduced to "Lineland," the single dimension of inhabitants knowing only north and south. The populace of Lineland consists of lines and points moving in a single direction, and each can only see the point immediately in front of or behind it. Women, in Lineland, are the points, and the lines, of varying lengths, are males with two different voices on either end. One cannot determine, except by the voices, which end is which; nevertheless those two voices are used by the lines to attract two wives, each of whom has a voice that harmonizes with one end of the line or the other. The distance between those two voices is also used in determining the length of each line.

A Square attempts to introduce the concepts of left and right to the "Monarch" of Lineland as well as the absurdity of lack of feeling and closeness in mating practices, all to no avail. This dream encounter with the ignorance of one-dimensional existence, the ignorance of men who discount all voices other than those that fulfill the needs of one end or the other, serves to prepare A Square for the visitor from "Spaceland."

Another preparation is the question his grandson asks during his math lesson with A Square on the last day of 1999, the query of a child unencumbered by the indoctrination of the rigid societal structure of Flatland. The question is a matter of thought progression: If one can square a figure, then what would happen if one cubed a figure and raised the power and the structure to another dimension? A Square, firmly enmeshed in the belief system of Flatland, is shocked and duly reprimands his grandson.

On that eve of the new millennium, A Square, in ruminating about the year just passing, including his grandson's question, says aloud, "The boy is a fool." A Square then feels a disconcerting presence, and a voice comes, as if from nowhere, saying that the child is correct. It is a visitor from Spaceland.

A Square, like any normal square, limited to concepts of left and right, north and south, has no context in which to see up or depth to discern a solid object from the third dimension (really the fourth, according to the author). He sees only a slice of that spatial figure, the slice in his own line of vision. Gradually, as the Sphere from Spaceland continues his entrance into Flatland, his initial appearance as a line becomes that of a circle from A Square's perspective, coming, as the voice did, seemingly out of nowhere.

A Square's astonishment is only the beginning of the trials and tribulations that accompany his introduction to Other Worlds. His fear, doubt, and denial escalate to anger as he defends his own world by attacking the intruder. Then A Square experiences a most extraordinary feeling, the pain of being touched inside, for the Sphere has, indeed, touched him.

The Sphere attempts to describe a higher dimension from which he came. He attempts to teach the concepts that expand This World into Other Worlds, particularly what Flatland is like from his perspective in Space where he can see inside (with his inside eye) the things that Flatlanders consider closed. Words are not enough to convince A Square, so the teacher resorts to deeds and lifts A Square to Spaceland. A Square describes his initial experience of space:

> "An unspeakable horror seized me. There was a darkness, then a dizzy, sickening sensation of sight that was not like seeing; I saw a Line that was not a Line; Space that was not Space: I was myself, and not myself. When I could find a voice, I shrieked aloud in agony: Either this is madness or it is Hell."

As he slowly adapts to sight in Spaceland, A Square is seized by a religious experience and perceives the Sphere as God. It takes some time before he realizes that the Sphere is not a god, and the Sphere informs A Square that he can only adjust to Spaceland by degrees. The Sphere urges A Square to return with him to This World, and in that visit back, after having experienced Space, A Square begins to discern the shallowness of Flatland and his former blindness.

Once again A Square is lifted to Spaceland where, with his newfound capacity "to see," he becomes enamored with his own superiority. His teacher informs him even "pickpockets and cutthroats" in Spaceland have the ability to see. The Sphere attempts to teach him more by his question, "Does this omnivdence [omnipotence] make you more just, more merciful, less selfish, more loving?" To which A Square replies in shock, "More merciful, more loving, but these are the qualities of women…" A Square is then informed that the best and wisest in Spaceland think more of affection than of understanding, more of the despised straight lines than of the esteemed circles.

Subsequent to A Square's acceptance of the third dimension (really the fourth, according to the author), through his experience of space, he is exhilarated with a new idea of thought progression, like his grandson, and entertains the notion of even higher dimensions. He envisions a dimension where four-dimensional existence would appear as ignorant and absurd as Lineland did to a Flatlander, as shallow as Flatland is to a Spacelander. Oddly, the Spacelander is unwilling to listen to such prattle and returns A Square to Flatland with a bump.

After his return to Flatland, A Square has another dream in which the Spacelander ushers him to the lowest depth of existence—to the nonmoving, non-dimensional, "vile and ignorant" land of the "Point." The single inhabitant of the land of the Point is completely self-contained, self-satisfied, and oblivious of any dimension other than his own; he attributes all existence to his own thought, unable to see, to hear, to move, or to change. A Square is astounded at such complacency and, try as he might, he is unable to budge him. The Sphere stops A Square's futile efforts with the words: "There is nothing that you or I can do to rescue him from his self-satisfaction."

During that dream visit, his teacher acknowledges his own error in silencing A Square's idea of higher dimensions. He then proceeds to initiate those dimensions "strictly according to analogy."

Convincing fellow Flatlanders of what he learned of Other Worlds is another dilemma for A Square. His efforts to enlighten serve only to make him an outcast and confined to prison in Flatland. In prison he begins to doubt himself, for there is no one of This World to validate his memories, his dreams, and his experiences. He ends his "memoir" with the hope that, in some future time, his words "may find their way to the minds of humanity."

It is my hope the following will not only validate but also bring life to Abbot's words.

## III

## Flatland, Spaceland, and Beyond to the Immeasurable

It has been known for two centuries that the left brain is the primary source of language. It was a connection made when damage to the left hemisphere through strokes or other brain injuries resulted in loss of the ability to speak. I have little doubt that Abbott was aware of that fact. Abbott's paper world, the written words and two-dimensional geometric pictures, suggests he was aware that language, plane geometry, and mathematics are produced by the same hemisphere. It would be over half a century before the specialties of our right hemisphere, Abbot's Spaceland, would be identified; primarily through the Nobel Prize winning work of neuro-biologist, Roger Wilcott Sperry.

We have a contra-lateral arrangement: the left half of the brain controls the right half of the body, and the right half of the brain controls the left half of the body. The left brain processes are considered linear, and the right processes are considered spatial, two different but parallel ways of knowing. The two hemispheres are connected and communicate, mainly, through a structure known as the corpus callosum. In rare instances left handedness may switch the higher cognitive functioning of those hemispheres. We know now, also, that the amazing human brain has the capacity to mend itself and reassign tasks to different areas.

Linearity is the principal mode of mental functioning in our society, which is the Flatland world Abbot portrays. Betty Edwards, in her book, *Drawing on the Right Side of the Brain*,[10] presents a brief but clear and informative synopsis of the history of societal prejudices for linear functioning and bias against spatial holistic processing, certainly true in 19th century England and an inclination that continues. She tells us it is a bias, embedded in the language we use, which favors group conformity over individuality, and that idea is expressed even in our political vocabulary.

Edwards says, "The political right (linear left brain), for instance, admires national power, is conservative, and resists change. Conversely, the political left (spatial right brain), admires individual autonomy and promotes change, even radical change. At their extremes, the political right is fascist, the political left is anarchist". Although I resist categorical statements (there are numerous human factors involved), I certainly acknowledge the basic relevance of her words.

Edwards' well-documented chapter titled, "Your Brain, the Right and Left of it," provides many intriguing examples of spatial and linear brain functioning. Edwards also tells us "the general view which prevailed (from the 19th century) until fairly recently, was that the right half of the brain was less advanced, less evolved than the left half—a mute twin with lower-level capabilities, directed and carried along by the verbal left hemisphere." But the split-brain phenomenon goes much deeper, as Abbott portrays in Flatland, and this prejudice for linear thinking has maintained This World and the fog that limits sensate existence.

As in linear Flatland, we often see the way we are trained to see. We feel, speak, and believe the way we are taught. The primacy of behavioral conditioning—punishment and reward—maintains the hierarchical societal structure where people continue to be treated as objects and their worth measured by conformity and successes in the objective world. Those feelings that are truly ours as human beings are suppressed, e.g., a man isn't supposed to cry or show fear, a woman is condemned for any expression of anger. It is also well known that prejudice, with its accompanying hatred, is taught. The words we need to speak are condemned to silence. Patriarchal beliefs lock the door.

But those unexpressed feelings and thoughts do not go away. They wind up as psychological baggage we carry in memory, and for some of us, the baggage becomes even heavier with an additional load of guilt. The oppression of our authentic selves is the source of mental turmoil. That emotional disturbance, which can take many forms, has also resulted in the use of anti-depressants, anti-anxiety drugs, and other psychotropic medications which use has reached shocking numbers. More often than not, the underlying cause is not chemical.

In This World of the present, women continue to be oppressed and abused, even in my former profession of psychology. We continue to analyze and label the irregulars—those who do not fit with our logical, sequential, and timely Flatland—under the umbrella of mental illness or convicted criminals who comprise the largest prison population per capita of any country in the world. There are many whose incarceration has nothing to do with justice.

Lineland exists in reality, for sex remains a prevalent focus of the male gender. It is the lowest common denominator by which men measure themselves. Half of marriages in the U.S. fail; many of the other half falter. Some of the causes, as in Lineland, are the lack of equality, communication, and closeness in relationships. There is a general consensus that the domination of men over women, that includes infidelity, is a justifiable underlying dimension of patriarchy.

The patriarchal belief system permeates all of our society. To state the obvious, it was even written in the Declaration of Independence, "All men are created equal," despite the fact that it is a country built on the backs of slaves and immigrants, not to speak of the genocide of the native population. Women are not even mentioned. Only through the work of brave women, no longer willing to be diminished in silence, has progress been made. That development has occurred despite the opinions of such people as Joseph P. Bradley, Supreme Court Justice. In 1872 he wrote: "The paramount destiny and mission of woman are to fulfill the noble and benign offices of wife and mother. This is the law of the Creator." After 72 years of concerted effort, begun in 1848, "suffrage," to have a voice, was granted in 1920 with the right to vote.

Advancement for the rights of women has continued in the past century, and some women have even been able to find niches in which they could rise to the status of men. Many have done so by continuing the "oh peace" (appease) cry, adopting the rhetoric of the high priests of our various subjects, maintaining, and even promoting patriarchy. Likewise, other oppressed populations, e.g. some of African-American descent, through the long and arduous battle for equality, have also been able to rise in status by compliance with the hierarchical structure founded in patriarchy, many of whom also learned appeasing rhetoric. In 2008, a man with mixed ethnic heritage achieved the highest elected office of our nation.

The battles for equality, though, have not been won; the double standards set up for women and men, and for those deemed to be irregulars, continue. Our country, like Abbot's Flatland, continues to be a self-governing, belief-based entity that is ruled by both seduction and fear. Despite our front of democracy (that is not democratic), and our engagement in world affairs (globalization that includes imperialistic motives), we have become a country that has many characteristics of a system that is closed. Laurence W. Britt makes a case for our extremist right politics and our deterioration to fascism under the previous right wing administration.[11]

The problem with a closed system is that it has a destructive impact both within and beyond its borders. While the political winds have changed for now, *Fear* politics persist. We will continue to swing backward and forward, politically, between the left and the right, for we have not learned from the mistakes of our past. And, as Abbot suggests, there are also "pickpockets and cutthroats" in Spaceland. The true cause is that the collective unifying force has been relegated to darkness.

Yet the spatial "knowing" brain capacity **is** a threat to society's standards of measurement, for it lends transparency to the shallowness and blindness of Flatland. The Other World that is holistic—seeing the big picture, not fragments. It is the world of intuition, imagination, and creativity, where time is irrelevant and time travel can be conceived. It is the world of symbols,[12] the language of dreams. It is the world of insight, the "inside eye."

The term used in the field of psychology for the "knowing which we do not know we know" is the *unconscious*, and, true to the bias of our cultural systems, we have been led to believe it is beneath our linear world, a menacing place of mythological demons. But Abbot, like Plato knew (as reported in his "cave" metaphor), and as I know now, it is the Other World which can lead us upward and out of the cave to the light. While our timely brain is necessary to keep us grounded, it is our spatial brain, working in symmetry which evolves us to higher levels of consciousness.

In the 19th century it had not yet been discovered that the fourth dimension is time. Perhaps Abbot knew it with his "inside eye" and displayed his knowledge by projecting his story forward to the next millennium. While we can explore our history in rational mode, looking backward in time, the spatial brain can look forward, as demonstrated by the phenomenon of art preceding discoveries in science.

On an individual level, the term used to describe dreams that foretell events or the dream-like production of *My Golden Frog* is *precognition*. It goes beyond intuition and enters the realm of visions, the metaphysical, considered by Flatlanders as *paranormal*, a word with negative implications akin to heresy. It is a more normal phenomenon than is generally admitted. One famous case is reported by psychiatrist, Dr. Diane Hennacy Powell:

> "Lincoln had a very vivid dream of walking around the White House and hearing all these people mourning and asking, "What's going on?" and then having someone tell him, "The president's dead." Then he saw his own corpse. He had this dream literally ten days before he was assassinated. He didn't tell anybody about it at first, but a few days before [his assassination], he told his wife and some friends."[13]

Rosemary Ellen Guiley[14] describes precognition as occurring most often in dreams but also occurring in waking visions, auditory hallucinations, flashing thoughts entering the mind, and the sense of "knowing". Guiley tells us what makes precognition so difficult to understand is that free will can change the perceived future as in the many incidents of individuals saving their lives and escaping disasters by changing their previously formed plans based on precognitive information. She reasons that if precognition is a glimpse of the true or real future, then the effects are witnessed before the causes, a condition that does occur in quantum physics. Guiley relates that "The most popular theory holds that precognition is a glimpse of a possible future that is based upon present conditions and existing information, and which may be altered depending upon acts of free will."

Guiley also makes distinctions between precognition, premonition, and prophecy: "Precognition generally involves knowledge of a future event while premonition involves the sense or feeling that something is going to happen; whereas all prophecy is precognition, not all precognition is prophecy."

It is difficult for me not to wonder if Edwin A. Abbot's work was prophetic and perhaps, in speaking clearly to our own day, his hope that his words will "reach the minds of humanity" could possibly be fulfilled.

Information regarding intuition, precognition, and the like was not available to me through much of my life, though, and prophecy was relegated only to the biblical distant past. Through those many years, my sensations of other dimensions had been suppressed with the belief that I was insane—a Spacelander in Flatland.

I had learned to adapt, and my life had become like the life of Abbot's narrator, A Square, who had risen to the status of the professional class through education. Like many women, though, I had done so by adopting the voices of Flatland, especially the "Oh Peace" (appease) cry.

Unlike A Square, my rise to space was from Lineland where a psychologist, who believed women have no memory, much less the gift of sight, disregarded my feelings and my thoughts. Over a period of eighteen months he reduced me to a mere dot on a line, and even that dot had been fractured. Lineland continues to be a profane place where women are violated, both literally and figuratively. There was no place to go but up—or die.

My first encounter with Other Worlds was a dream of Spaceland where I could see, dimly, that I was boxed in by structures of measurements built by men; perhaps because a line or a dot cannot rise in any dimension.

In This World the universal godhead remains exclusively male; women continue to be non-existent in that godhead except for her role in reproduction. The high priests of our various subjects continue to be dominated by men. Even in the creative fields of art and music, there have been few female artists and no major female composers who have gained recognition. It is not that women are not capable, we are afforded little opportunity given the time consumption of *womanly* tasks—the "benign offices of wife and mother."

Back to the dream which was a most extraordinary experience. It, too, provided images of where I was, where I had been, and a prediction of the future, something I did not understand at the time. The shape of the enclosure and some of the symbols suggested that it was a box-like hall in which I was trapped. The furnishings were symbolic of my familial history and the history of generations before me. The dream produced the peculiar sensation of me being in the center where all of the sides of that structure were visible simultaneously as though I had 360 degree sight. I also sensed there were other dimensions that were not visible.

One of the dream symbols was an oval mirror, shaped like an eye, hanging on the wall, but most of the mirror was blocked by a square piece of wood attached over it. The only apparent means of escape from the dark enclosure was through a frightening, shadowy, and veiled doorway.

The highlighted symbol in that dream room was one I did not understand. It was unsettling, for I knew it had something to do with my beloved daughter.

In my first actual encounter with Spaceland, 18 months later, I did not have a guide or forewarning and was lifted up and thrown blindly into space where the darkness evoked the anguished words, "My Soul is Black." Then a figure took form in a picture I drew. It was an image so frightening even the strongest would feel nausea. It was a ghastly figure with empty holes where eyes should have been, the mouth, the opening of a vagina. The anatomy of the vulva was constructed by a highlighted image of the popularized carved-in-stone figure of male praying (preying) hands. I lived in absolute terror for days.

Gradually the fear subsided, providing a short-lived sense of equilibrium. A week later, during a moment of extraordinary tranquility, the box of my Spaceland dream reappeared in my mind. Everything was identical except the block over the mirror had miraculously been removed. I use the word *miraculous*, for what I saw next was transforming:

*Through the mirror which was not a mirror, I saw a light that began to move and flow in every possible soft color, in dimensions that even defy imagination and certainly any concept I have of space, of direction, of size, or shape. In that vision the structure of the box totally dissipated, as did its contents. Everything seemed to be turning inside out and upside down to what it was and absolutely topsy-turvy to anything rational.*

Awe is the only word approaching a description of the sensation the vision evoked in me, and it has taken these many years to fathom what I saw that August 6, 1991—the "I" behind insight. It is not a dimension that has a direction yet encompasses all direction. It is the unifying force that exists throughout the living universe holding everything together in one great symphony. I can still hear the voice in my head that at the time had no audible sound: "I can see everything!"

Those words were thick with meaning, for there was a literal change in my vision and perception. I saw the smallness of my Yorkshire terrier for the first time. I could also see across time and space, an experience I described in my journal:

*While sitting with a group of friends at a café the next week, attempting to engage in the conversation of the moment and appear normal, I looked across the street at the church I had seen many times before. This time, it was a different seeing. I saw not only the spatial form of the building but the history and evolution of architecture. I saw the actual construction of the structure beginning with the foundation, then the walls, and the roof. The history of the thousands of years of belief that went into the building of that edifice came pouring into my mind, both the destructive and constructive forces represented there. I had flashes of memory, like a film running at high speed, the oppressive and inhumane abuses by the church "in the name of God," and the feelings evoked were chaotic.*

It was an awakening to the shallowness of Flatland and my former blindness.

Like A Square, I was also introduced to the "vile and ignorant" land of complacency, not by a dream, but in reality. It is visible in glass houses on the best side of town. It is written in the language of flatlanders, words like *turn the other cheek*. It was produced by the tongue of our former president when Bush said, "I don't spend much time looking in the mirror." I have heard it out of the mouth of a man I once admired, a new-age Flatlander: "Bad things do not happen." And the words from an acquaintance even more astounding: "Evil does not exist." I confess to repeating mindless ideas in my past, particularly, "All things happen for a reason." These are symptoms of three-monkey thinking—deaf, dumb, and blind—that helps to keep Flatland flat.

To refute those complacent words: our earth is a delicately balanced living body, both accidental and orderly. Bad things do happen and to both bad and good people; the sun shines on everyone, not a chosen few. Many of the *natural* disasters occurring on our planet, though, are not so natural but caused by disregard for our life-sustaining earth. And the human waste—the human suffering—is also most often inflicted by fellow humans.

Like A Square, I, too, have been introduced to other dimensions, strictly by analogy. Theoretical physicists[15] are now proposing parallel dimensions and/or universes. There is much conjecture about those dimensions as in the PBS TV series, *The Universe*, and millions of sites on the Internet.

Closely related to parallel dimensions is another phenomenon known as *synchronicity*, coincidences that are more than coincidental. It is a word introduced and used by Carl Jung. Its generally accepted meaning: the experience of two or more causally unrelated events occurring together in a meaningful manner. But Jung attempted to explain it in pseudo-causal terms as a manifestation of the unconscious and a "collective unconscious" (cultural bias). A. N. Whitehead, in *Nature and Life*,[16] considered such theories as "misplaced concreteness." Arthur Koestler[17] elaborates that *misplaced concreteness* as analogous to theologians who start with the premise that the mind of God is beyond human understanding then proceed to explain how the mind of God works.

In that same line of thought, precognition, parallel dimensions, synchronicity, and Spaceland encounters give rise to many other conjectures such as reincarnation, predestination, and even aliens. They are explanations used to assuage mystery, for mystery is a threat to illusions of control in both Flatland and Spaceland. Such explanations give rise to myth. The randomness that coexists with an underlying order and harmony in nature **is**, at our present stage of evolution, beyond our comprehension.

It was simply mystery when my golden frog appeared, and "my two eyes, seemingly open to separate worlds," did arouse a dissonance in my mind. It was a being so exquisitely visible, tangible, but even more, it was an *experience* of the harmony which permeates all existence. That its appearance occurred on a day very special to me, adds the element of synchronicity.

That frog was a symbol which, like all symbols, had multiple layers of meaning. Radical evolution in a single lifetime was among them. I have experienced other equally intense encounters with nature that I will reveal in time. Intuition, though, shows up for me more frequently.

Like the day I had a nagging feeling something was missing in Abbot's narrative, for it seemed as though my own story had strayed from *Flatland*'s metaphor. I went to my computer and checked it out on the Internet. There it was in the 2$^{nd}$ and revised edition of *Flatland, A Romance of Many Dimensions*, published in 1884, a preface by the author to clarify misunderstandings and criticisms regarding the first edition—that problem of interpretation.

In the preface he speaks of dimensions beyond the fourth. He argues that dimension implies direction and measurement and when we go beyond direction and measurement, we cannot know what to measure or in what direction. Yet he believes it to be there: "Even I cannot comprehend it, nor realize it by the sense of sight or by any process of reason; I can but apprehend it by faith." It is as though he is describing my inside-out, upside-down vision, the "**I**" behind insight, a glimpse into my very soul.

Abbott also elucidates his thoughts on patriarchal hierarchy: "those who maintain superiority over multitude of their countrymen by their intellectual power are in conflict with nature and nature has condemned them to ultimate failure." He states: "I see a fulfillment of the great law of all worlds, that while the wisdom of man thinks it is working one thing, the wisdom of nature constrains it to work another and quite a different and far better thing."

Abbott quotes a Spaceland poet: "One touch of Nature makes all worlds akin" to refute the human family's likeness that "runs through blind and persecuting humanity in all Dimensions! ... we are all liable to the same errors, all alike the Slaves of respective Dimensional prejudices."

Abbot also admonishes his readers, albeit in circuitous language, about becoming circles. He addresses "Spacelanders of moderate and modest minds who—speaking of that which is of the highest importance, but lies beyond experience—decline to say on the one hand 'This can never be,' and on the other hand 'It must needs be precisely thus, and we know all about it.'"

It is a warning regarding dogmatic thinking and simultaneously offers credibility to experience.

In Abbot's dedication to his revised edition, he challenges others to aspire to yet higher and higher dimensions, contributing to the enlargement of the imagination and the development of that "most rare and excellent gift"—modesty.

## IV

## The World is Mostly Flat

Modesty was not one of my concerns for many years after my Spaceland encounter; it would have required a mighty climb for me to reach that level. It seemed as though I, too, was confined to prison in Flatland. My so-called therapist had also shattered any semblance of self-esteem.

That fracturing was part of a vortex in my life, for shortly afterward a long-term client died tragically. There was little breathing space before the death of my father, March 8, 1992, and the next month, an uncle died. The appearance of my golden frog on January 28, 1992, the day after my brother's birthday, was a miracle during that darkness and stilled a major life-long fear: I knew then I was not insane. It also gave me hope to overcome the sensation of drowning.

In June 1992 I visited my companion and former husband, a major player in my story, in Bialystok, Poland, where he was working at the time. It was intended to be a vacation, but there was no respite from the process that had begun. The day of my arrival, during a daytime rest, I had the worst dream of my life, at least the only one from which I was awakened by the sound of my own screams. It was a dream reiterating the highlighted symbol in my "box" dream: I was carrying my boxed-in inheritance from my family of origin to another generation, to my beloved daughter.

That is the way with dreams. If one does not comprehend or attend the message from his/her spatial brain the first time, it will be repeated in a more emphatic form. That is, of course, unless one does not give credence to dreams, dismisses them, or suppresses recall for any reason. It is a fact that everyone dreams. At the time, though, I had come to welcome messages from my unconscious, no matter their content.

Two other significant dreams occurred during the first week of my visit. One of those dreams was:

*Listening to a large older man giving a lecture in psychology; I knew it was Jung, and he was saying that he disagreed with my assertions. Then he walked away, and I knew I had banished him.*

There was considerable background to the dream. Since my Spaceland encounter and the destructive experience with the therapist, I had been rereading psychological theories in hopes of finding some explanation for what had happened to me.

I am in agreement with many of Carl Jung's concepts, especially his "coming of the unconscious" idea discovered during my research. In his explanation of that phenomenon, though, his subjective confession becomes apparent with his obsession with "archetypal" myth.[18] Myths are myths and are reflective of a culture; they are also reflective of the "fog" or prejudice of that culture.

Archetypes are another story. Those recurring patterns are evident in our genes that dictate the formation of our bodies and a vast array of instincts, basic endowments existing in everything that is alive.

I had found additional data regarding Jung in Phyllis Chesler's *Women and Madness*.[19] It was Jung's opinion regarding women: "But no one can evade the fact, that in taking up a masculine calling, studying, and working in a man's way, woman is doing something not wholly in agreement with, if not directly injurious to, her feminine nature."

It was just one more nail in the box, just one of patriarchal psychology's countless contributions used to oppress women. The dream of Jung was exposing to consciousness my banishment of all the men and women of psychological fame who believe they know the thoughts, feelings, needs, or true capacity of a woman; or anyone else, for that matter.

The next night I dreamed again, one that seemed to be providing guidance: *I was standing, pointing, and saying, "Popacz!"* (*Look!* in Polish)

And I did.

The following week, on one of our excursions exploring his homeland, my companion drove twenty-five miles to the border of Belyrus, formerly part of the USSR. At the border crossing—a narrow bridge blocked by a rusty chain and a stop sign—he turned onto an unpaved road to follow the border. The road passed through a small community of farmers. I particularly recall seeing a woman sitting in front of her wood house, her eyes following, intently, the strange car that was passing. Suddenly an incredible sensation of loneliness and despair washed over me. I breathed deeply and forced myself to focus on the road we were travelling.

After passing through the tiny community into open country, I asked my companion, "Where is the border now?" He pointed and said, "Over there." I asked him to stop and show me. He parked the car beside the dirt road, and we got out.

All I could see was the sun shining on an open field with a gradual slope to a stream meandering in the valley below. It was a moving, teeming scene, impossible to capture on canvas: blue flowers mixing with the greens, yellows, and reds of foliage, soft colors seeming to flow in the gentle breeze and in the bathing light that knows no borders.

"Exactly where is the demarcation line?" I asked in disbelief, for my thoughts were of the *Iron Curtain*, words that painted images in my mind. There was nothing there, not even a fence. Then my companion told me of a plowed strip of earth that had split and divided people, a strip that had seeded and mended itself. My mind was racing, and I asked, "How could they prevent people from just walking across the line?" and his answer was, "Fear." It was a fear, he explained, with foundation, for many were killed or imprisoned for crossing the "Iron Curtain." It is a myth that "There is nothing to fear but fear itself."[20]

To borrow the words of Abraham Maslow, one of the founders of humanistic psychology, it was a *peak experience*—another touch of nature—a major insight. Many liken such peak experiences, similar to A Square's encounter with Spaceland, to a "religious experience."

On the contrary for me, my sudden insight was that the borders in my mind, the borders that had split and divided me, were not natural but made by the closed systems men have invented, especially religion and the *religion* of psychology. It was a moment of recognition—the chains that bind the mind[21] are forged by belief and fastened by fear. It was also a moment of decision: to face all of my fears, regardless of the cost. The words that came to my mind were contained in the last verse of Longfellow's poem:

> Let us then be up and doing
> With a heart for any fate,
> Still achieving, still pursuing,
> Learn to labor and to wait.

I sensed Poland was where I needed to be at the time. My intuition was that I could no longer be a support to anyone, including myself. Another thought encouraging me to *look* at Poland—it is almost impossible to see a system while enmeshed in the system. One can gain a clearer perspective from a distance of either time or space. I needed that distance in order to see the systems in which I was enmeshed in the U.S.

I was invited to provide lectures at the University of Warsaw, Extension in Bialystok. My companion voiced his decision to follow his feelings and asked me to live with him.

It was almost magical the way things just fell into place. Within three months after my return to my home in Maitland, Florida, my house and office were rented, clients transferred to new therapists, my pets transported to my daughter's house where I also stored personal and professional belongings. I left the U.S. with two large travel bags: one filled with daily essentials and clothing, the other filled with psychology and science books. My so-called *choice* of books was no less magical.

༄

The gift of curiosity has been with me throughout my life. Even as a teenager I longed to travel, to see what was beyond the horizon. After graduating from college at the age of 40, I took my first of many trips to Europe. It is truly a broadening experience to travel; it is quite a different experience to live in another country.

I had a special attachment to Poland. I had seen its bleakness in 1985 while still under Communist rule. I had seen the scarcity: long lines to buy the bare necessities, black markets to put food on the table or to find gasoline for an automobile if one were lucky enough to own one, and the hoarding of a simple thing like toilet paper. I had a sense of familiarity with that desolation and poverty.

Yet the miracle of the human potential was there in Poland. It was the capacity for people to work together in times of disaster which enabled the rebuilding of Warsaw out of destruction, a people who survived, at least some of them.

During my 1985 visit, I explored the Holocaust Museum in the reconstructed Old Town Section of Warsaw which documents scenes one does not readily forget. The stories of the annihilation of countless men, women, and children were depicted in photographs of skeletal humans, mounds of rotting flesh and bones. I, along with the other visitors, kept the silence of that tomb-like memoriam, for no words can portray the horrors of what happened there. Poland aroused my sympathy.

I have learned, though, that what things appear to be on the outside has little to do with what is going on inside. I did not realize at the time that a closed system is a closed system, whether it is a person, a family, a region, a country, or any of its subsystems.

More of my long-held beliefs were shattered during my nine-month stay in Poland, some by my own countrymen. I saw many instances of the abuse of human freedom. Grant money intended for the welfare of the Polish people did not reach its destination; used up instead by U.S. corporations' management fees. Interest on World Bank loans were charged at 50 percent. These are crimes for which there are no laws, and no one seems to be accountable. Then there were the investors ready to pounce and prey on anything of value. Poland had been a country closed for half a century by Communism. The Polish people did not talk about the fact that it was a country handed over to Stalin, whose capacity for inhumanity easily rivaled Hitler's, by the not-so-wise leaders of England and the United States after World War II.

You will not find in the history books of Poland what I discovered while living there. Poland is another Flatland where the patriarchal father reigns supreme by the indoctrination of its fundamentalist

Catholic Church, the pinnacle of hypocrisy and sexism. It is a place where St. Mary is idolized, as are women, at least in posturing, for men bend to kiss their hands. Yet I could not walk down the street without noticing women with black eyes and bruises. I lost count of the number of sexually abused encountered, both women and men, by the *fathers* of the church. The abuse of alcohol was evident on the streets and in many of the homes I visited. And depression seemed prevalent.

Poland also has a caste society that changes by degrees from the high priests of all the subjects diminishing to the lower castes that includes farmers. At one time there had been an even lower caste: the Jews. There were no Jews in or around Bialystok; the 25,000 remaining in Poland, which was once the homeland for the largest Jewish population in Europe, live in the more urban areas. The extent of the prejudice is revealed by Jan T. Gross, a Professor of Politics and European studies, in his controversial book, *Neighbors: The Destruction of the Jewish Community in Jedwabne, Poland*.[22] I visited the site, 40 miles west of Bialystok, and was aware of the massacre, but it was portrayed as the work of Nazis. Gross reveals it was at the hands of Polish people. I found Gross' report plausible after encountering the anti-Semitism that continues.

As in any closed system, disease runs rampant. In Poland it was and is the disease of hypocrisy and the fog of prejudice. That bigotry, fueled by generations of Christian brainwashing, resulted in 1600 men, women, and children being "clubbed, drowned, gutted, and burned ... by people whose features and names they knew well: their former schoolmates and those who sold them food, bought their milk, and chatted with them in the street." There was also the monetary gain, for the properties and possessions of the dead were taken over by Poles.

There were some heroes—a family known to have saved surviving Jews—but they were derided and driven from the area. Jedwabne was only one of three Jewish Communities in the region that met the same fate. Those who did not participate in the massacres looked the other way.

A line from one of my poems is appropriate: "Silent, killing evil does exist in the eyes of complacency."

The similarities of rural Poland in 1993 to rural central Florida where I was born a half-century before were astounding. To my knowledge there were no Jews in Linden, but the prejudice against blacks, gays, northerners, Catholics, anyone who was different, was unquestionable. Frequently I felt as though I had stepped backward in time. Scenes stimulated personal memories, like passing the wood house near the border crossing when the incredible sensation of isolation and loneliness washed over me. The house was similar to the converted log cabin, the homestead of my childhood. Isolated—seemingly in the middle of nowhere—where my curious, hopeful eyes watched people passing on the dirt road in front of the house.

There was also something about being in a place where I did not speak the language that played a role in arousing childhood memories, a time when "children should be seen and not heard" (which saying was originally applicable to females). Memories taking form in images of what it was like to have no means of communicating what was going on inside of me, reminiscent of another dark period in my life, as though this one was a repeat performance. With the loss of "senses learned," my authentic feelings had begun to surface.

I particularly recall a Sunday morning when I saw through the kitchen window, a family riding in a wagon pulled by a tractor, no doubt going to church. A little girl sat in the back, and her face appeared to be as fixed as a frozen rose. A stoic face on one so young was appalling. Yet it reminded me of numerous scenes of my own life on the family farm, particularly a Sunday when the dirt road to the highway was flooded, and I rode on the back of a tractor to Church. I had a sense of my own emotionless demeanor with so much going on inside.

Those memories, along with numerous others, though, did not seem to interfere with my capacity to think, to see, and to make connections as I continued my search for some explanation of all that had happened to me. The protestant work ethic (also known as workaholic), deeply engrained into my life, began to subside when I realized what a precious gift my companion gave to me during my nine month stay in Poland: time. Time to pore through the books I imported, searching for what, I did not know.

The most informative Psychology book was a gift from a friend before I left the U.S.: Henri F. Ellenberger's *Discovery of the Unconscious, The History and Evolution of Dynamic Psychiatry*.[23] It is a well-documented, lengthy volume, the result of two decades of work. In that book Ellenberger examines the many dimensions involved in the formation of psychological theory: socio-economic, political, and cultural backgrounds, as well as the personality of each of the pioneers and their personal environments.

Ellenberger traces the theories of Pierre Janet, Sigmund Freud, Carl Jung, and Alfred Adler. Janet was one of the first people to coin the word *subconscious*. He also drew a connection between historical traumatic events in the subject's life and his or her *hysteria*. Alfred Adler was also aware of relational impact on patients and believed no one can be truly seen outside the system in which they live. He was the first community-minded theorist whose emphasis was social justice.

Freud and Jung, however, suggest people are, by nature, closed systems, and the problems encountered are born in the psyche. Their theories also suggest that the source of evil is within, as in Freud's *id* and Jung's *shadow*. Jung was even more specific when he said that children's conscious problems are a result of "the evil within oneself as well as outside."[24]

There is no question that Dynamic Psychiatry has changed since the founding fathers' theories. The term *evolution*, though, is not necessarily appropriate, for the changes have continued to be an abstract or a reflection of the socio-economic, political, and cultural backgrounds, as well as the personality of the theorists and their personal environments. Psychoanalytic psychology/psychiatry has fallen in and out of favor with the fads, the cultural or politically correct view of the time.

Ellenberger's book evoked many thoughts about the unconscious, especially about the dreams that initiated my Spaceland encounter and how my dreams were leading me toward transformation long before meeting the so-called therapist. I became firmly convinced that the purpose of dreams is to bring synthesis to our Selves and to evolve conscious awareness. It dawned on me that dreams are also a part of the creative process, and symbols are formed from one's subjective world. The dreamer is the **only one** who can know the meanings of their own symbols. Those dream symbols are also evoked from the context of a person's life, usually what is going on at the time.

The phenomenon that clients frequently present a dream predicting the course and outcome of therapy was not something I had been taught but an incidence stumbled on in my research. My "box dream" occurred two weeks after my second appointment with the so-called therapist. The block over the mirror was removed after I banished him from my life.

Knowing of that phenomenon may not have changed the outcome; my dream comprehension had been contaminated not only by the therapist's flawed analysis, but by Freudian and Jungian imagery in the dream dictionary I had read. I had not trusted my intuition regarding the meaning of my dream symbols. I had not trusted myself, in general, which contributed to my falling prey to the psychologist and numerous other *authorities* in my life.

I began to contemplate all of the sub-divisions in psychology, those various denominations, so to speak, that do not give credence to the unconscious or even dreaming and instead treat those numerous obsessions—the symptoms—as the disease. It became obvious that psychology is similar to religion with its variations of dogma.

I revisited my "box" dream which contained so many symbols that, elaborating each of them, individually, with their multiple layers of meaning, would fill an entire book. Some of the dream symbols elucidated, precisely, what obsessions are: attempts to escape the baggage we carry—disowned feelings and thoughts—as well as escape from the reality of the here and now. Obsessions take many forms: drugs, alcohol, food, work, clean, money, sex, religion, other people, art.... The list is endless.

Even then I sensed my own obsessions—where I used up my energies—were simply cover-ups to evade grief and an underlying basic need to be seen, heard, and acknowledged. My need for a mirror grew into an insidious dependency. Without my spatial brain activity, obstructed by Flatland conditioning, I spent much of my life searching for myself in the eyes and words of other people.

That search was connected to the terminology I had used, of being *fractured* by the therapist. Now I know dependency essentially allows someone or something else to be one's mirror and makes one vulnerable to all sorts of distortions. Time and again the symbolism of *Humpty Dumpty* and even the words of that little poem rang true for me. I was able, finally, to interpret the symbolism in the context of a dream of a nuclear accident:

*I was driving a stripped-down car. Coming up behind blocked traffic, a policewoman stopped me and told me to wait because I was last, that she would be back. She said I would have to take my clothes off so I could be washed, and I sensed there had been a catastrophe. Then I noticed two buildings (that seemed to be Chernobyl) on the left and wondered about the nuclear accident.*

That my stress was showing up physically was not a doubt, nor was it a doubt that I was blocked from moving forward by the contamination of guilt imposed by self-blame. But I no longer wonder about the nuclear accident; it was not my center but my ego that had been fractured. I can now define *ego* as a set of stories about oneself adopted and believed. My beliefs about myself—a self-reliant, capable, and strong woman—achieved through a lifetime of accomplishments, had, indeed, been fractured.

The variations of *ego* within one's own life and among people in general have to do with many dimensions of interaction with one's environment, including the diverse ways one is treated or seen. Caretakers play a major role in shaping ego development. A child can be diminished in both subtle verbal ways and in profound ways: abuse, neglect, and abandonment, to name a few. A child can also be overindulged, grow up with a sense of entitlement, and never learn to take responsibility: an inflated ego. But our *egos*, the beliefs about ourselves, do not define our true selves, for it is a shallow and empty Flatland form of measurement.

Yet those egoists abound and become the lauded circles in Flatland's field of psychology. One particularly, Harvard Professor Steven Pinker, a specialist in language and cognition, addresses what is referred to in *My Golden Frog* as "mechanical needs" and my "robot program," two distinctly different ideas. While our genes preprogram our physical bodies and its functions (e.g., breathing, eating, walking on two legs), our robot programming is accomplished by Flatland training.

In Pinker's subjective confession of how his mind works (*How the Mind Works*[25]), he adheres to evolutionary psychology, clearly different from evolution of consciousness. His is a mechanistic computational model of the functions of the brain. His mind-as-computer representation is based on genetic programming of human behavior by natural selection. That mechanistic view suggests that we, as humans, are essentially problem solvers, another corrupted inference of creativity.

Other cognitive psychologists who, until 10 to 20 years ago, gave little credence to the unconscious, frequently favor genetic trait theories. Such models—biological determinism—have no explanation for free will, the fact that people can and do change, much less the creative process.

In his book Pinker speaks of the "selfish" gene which has come to be a way of expressing the gene-centered view of evolution. In subsequent writings, Pinker also promotes language as a reflection of the user's "nature" rather than the user being shaped by the cultural language of one's environment.

Another *genetic* program, just as ludicrous, is the proposal by another of Harvard's faculty, Martha Stout, in her book, *The Paranoia Switch*,[26] when she states that even political leanings—conservative or liberal—are partly "born in the blood." Stout relies on data acquired by a 2005 research questionnaire prepared by political scientists. Such questionnaires are frequently designed to achieve the bias of the researchers, whether conscious or not.

Stephen Jay Gould, Paleontologist and Evolutionary Biologist, who was opposed to evolutionary psychology, also had a lot to say about such "scientific" research in his book, *The Mismeasure of Man*.[27] He cited many cases of totally refuted Flatland measurements. One of the more infamous is an 18th century theory of phrenology, a theory stating that the personality traits of a person can be derived from the shape of the skull. That theory was the product of "rigorous scientific study," confirming "all the common prejudices of comfortable white males—that blacks, women, and poor people occupy their subordinate roles by the harsh dictates of nature." Gould described such researchers as "trapped by their rhetoric" and who "believe in their own objectivity." Gould went on to ask, regarding such so-called *scientific* data, "Shall we believe that science is different today simply because we share the cultural context of most practicing scientists and mistake its influence for objective truth?"

Gould explained that we are inextricably part of nature, and human uniqueness resides primarily in our brains; human societies change by cultural evolution, not as a result of biological alterations. He stated that "'nothing but' an animal is as fallacious a statement as 'created in God's own image.'"

There were two startling discoveries in Ellenberger's book addressing those same issues. One was in the form of a question Ellenberger asked about what was missing in all those dynamic theories he researched (and all other psychological theories)—the mytho-poetic function—the failure of those *fathers* of psychology to address creativity in all its mystery, a capacity that contributes to making us distinctly human.

The other discovery was Ellenberger's sentence which presents a twist in thought: "Man created god in his own image." Such a simple statement, yet it is a concept that explains the diversity of religious concepts of *God*. Religious doctrines, like psychological theories, are the product of the socio-economic, political and cultural backgrounds of its proponents, as well as the personalities of the pioneers and their personal environments.

I had already been questioning my Flatland training, but after those two discoveries, I began to wonder if any of my so-called education in psychology held truth. Every time I had questioned a theory, even in graduate school, I was met with derision and hostility. I began to see that not only are religion and psychology similar, both are an outgrowth of patriarchy, and both present double binds, essentially lose/lose situations.

To challenge or even question the dogma of religion, the word used is *heresy*, and the history of religion reveals the condemnation of heretics. While we are no longer burned at the stake or beheaded in the U.S., we are marginalized, ostracized. If we do not challenge, then we remain prisoners of Flatland.

Similarly, in the therapy room, questioning the perception or judgment of the psychologist is called *resistance*, which is considered to be indicative of pathology. Not to question results in dependence on the psychologist as *the* authority and makes one vulnerable to those perceptions, judgments, or beliefs.

I was not only facing some of my fears in challenging the *religion* of psychology, but traditional religious beliefs. I was moving more and more toward agreement with Nobel Laureate Bertrand Russell, a philosopher who believed in freedom of thought and humanitarian ideals. He maintained that organized religion, despite any positive effects it may have, "serves to impede knowledge, foster fear and dependency, and is responsible for much of the war, oppression, and misery of the world." Russell considered the ideology of Communism as effectively the same as religion. I can now add another dimension to that stance: Religions contribute to the silent, killing evil that does exist in the eyes of complacency.

As to my teaching experience, even though I was well prepared and well organized to present psychological theories, the presentation lacked cohesiveness. I was attempting to teach something I no longer believed. There was also interference by my Spaceland knowledge which was still in the formative stage. A colleague summed it up when he said, "You teach like a poet." Perhaps a way of saying my language was thick with underlying meaning and lacked clarity. I essentially lacked the skill of expressing my thoughts in language and was about as well received as any Spacelander in Flatland.

By the second semester I realized that *underlying meaning* was my own story in disguise, evidenced in the case studies presented, a subjective confession, as Jung suggested, from which there is no escape. Even in my attempts to get *me* out of the way, I was offering my own biography, for I began to teach by experiential methods in small interactive groups. I urged students to express themselves, their own stories and thoughts, in journals, games, drawings, anything I could devise to tap their spatial brain knowledge, things they knew but did not know they knew. I also attempted to teach them "to see" as they learned to work together.

I was faced with derision by Flatlanders, especially two American women, one a Teach America professional who personally hated journaling; the other a counseling psychologist whose subjective confession was revealed in her words: "People do not change".

There were some, though, who expressed new-gained insight from their experiences. Like a young man, quiet in class, who gingerly approached me on the street one day to express his concern and his appreciation. He said, "It is like waking up and seeing my own self for the first time, but it makes me wonder how much of my whole life has been and is unconscious." Or like another quite girl in class who wrote in her journal, "What an amazing thing we can build together."

The colleague who said I taught like a poet, summarized the subject of my teaching activities as *Symbolism and the Evolution of Consciousness*. It was the only instance in my process when I was lifted up and encouraged by a psychologist.

∽

When I wrote *My Golden Frog* on April 14, 1993, the experience of the writing and the words were startling. It was comparable to another major event that, together with my peak experience of discovering nature with no borders, was to change the course of my life. It can only be described as beyond otherworldly and one that was foreshadowed in the words, "remnants of human waste piled in chaotic heaps," which had more than a single meaning.

It occurred in mid-May 1993 when I went on another excursion with my companion to Lublin. I had heard of Mejdanek, a Nazi death camp museum near that city, and decided to go there on my own. My companion had refused my requests to visit Auschwitz.

Much to my surprise, Majdanek is not out in some hidden place but at the edge of the city with a wire fence dividing the death camp from a residential section. There is an iron barrier separating the site from the highway, an enlarged facsimile of the barbed wire that had imprisoned so many, for even when the camp was active, the Nazis did not bother to hide what was going on. Behind that iron façade is a wide expanse of concrete leading to concrete steps that rise toward the stone entrance.

Before entering, though, another set of steps suddenly sent me down into a pit-like area lined with more stones. I had to climb again.

As I walked up those steps, the pillars on both sides guided my eyes through the opening and down a long, paved path to a massive dome-shaped structure that seemed to rise on the horizon. As I walked that path, I passed a group of young men carrying flags with the Star of David imprinted in blue. Some of them were leaving the barracks where their people were housed. Then I saw other barrack-like structures with chimneys. Whether it was the cold, damp weather or the horror I saw with my own eyes, a chill crept through me when I approached the dome, for beneath that structure was a huge mound of ashes mixed with telltale fragments of bone.

There is a sign in the crematorium that reads, "One thousand a day," and there are ovens lined up and stacked to facilitate that goal. Even if Lublin's townspeople did not want to look at what was happening there, I wondered how they learned to live with the smell of burning human flesh.

I walked through all the rooms, and one I particularly recall was furnished with a stone operating table which evoked many images of horror and torture in the name of experimental medicine.

When it was time to leave, I began to walk toward the exit path but had to pass that mound of ashes again. Suddenly another weakening chill surged through my body, and I leaned against one of the supporting beams of the dome. Then the wind made an eerie sound that grew in intensity. It seemed to be carrying howling, moaning, crying voices, sounds of primal grief so deep, words cannot begin to touch it. I stood there, frozen, as though suspended in time, and then actually heard the sound of my own voice whisper in reply: "I will never be silent again."

That experience was incomprehensible by any process of reason, yet it was no less real. It was and continues to be simply mystery.

※

Throughout my scholastic life I have been more adept at math, mechanics, and organization than with language. During my high school years I wanted to become a physicist. Yet biology and anthropology were fascinating to me, and books like Peter Farb's *Humankind*,[28] tracing the evolution of humans, were appealing. That background, along with an undergraduate course I took in Physics and Human Affairs,[29] contributed to my turn to the natural sciences for answers.

Plagued with curiosity about my moving upside-down, inside-out vision, I had telephoned Dr. Banchoff, professor of mathematics at Brown University who specialized in the geometry of higher dimensions. He had no answers to my questions about what I saw. He referred me to his book, *Beyond the Third Dimension*[30] which referenced *Flatland*. It was one of the science books I took with me to Poland. The others were James Gleick's *Chaos*,[31] *A Brief History of Time*[32] by Stephen Hawking (a remarkable man), Theodore Schwenk's *Sensitive Chaos*,[33] and Leonard Shlain's *Art & Physics*.[34] Concurrently, as I was confronting and banishing much of psychological theory, I was studying all of those books and making extensive notes. I sensed a breakthrough one day while experimenting with math and charts, an attempt to figure out how time, as a fourth dimension, would affect our three dimensional construct of humans.

Two weeks before leaving Poland, while my companion was away in Russia, I began to cut and paste all of those notes together; then set to work to bridge those borrowed ideas into something more fluid. I would type for a couple of hours then fall asleep, only to awaken with dictation going on in my head. Over and over again my sleep-waking-dictation experience occurred until I had 14 pages of words so thick it took me years to understand them. It was truly a synthesis of my study, yet one truth holds: one can only adjust by degrees.

Those fourteen pages were based on a space-time-matter-energy continuum of humans, trashing the three-dimensional Flatland concept of measurable and definable humans-as-objects or psychology's divisions of the self. Those pages were an answer to a lifelong query about a higher plane of existence.

It is clearer to me now, and I can say it more succinctly: Humans are multidimensional yet indivisible, basically comprised of our spatial brains, our timely brains, our bodies, and the energy that is life. We have evolved to reflect, and to reflect on, the world around us in our spatial right brains, where time is irrelevant; in our timely left brains that provide a sense of structure; in our bodies (consisting of the multiple dimensions of biology: chemistry, mechanics of movement, proportion, genetics . . .). Then there is the mystery of energy that imbues matter with life. *Energy* is a word so bandied about, yet no one actually knows what life-giving energy is.

The word *continuum*, in regard to the multidimensional human, is essential in its established meaning: "a continuous extent, succession, or whole, no part of which can be distinguished from neighboring parts except by arbitrary division." Those arbitrary divisions abound.

A dream provided a holistic image for that 14-page manuscript. It was an image of the earth suspended in space as depicted by those remarkable photographs captured by astronauts. It was so validating to find, years later, that one of the great minds in human history, Leonardo da Vinci, in his world renowned drawing of the Vitruvian Man, was depicting the blend between art and science and relating man with nature.[35] I elaborated my dream with a female persona,[36] for no-one speaks of *father* nature.

Leaving Poland June 14, 1993, was painful, for I left my companion behind. I was supported though, by that world image, a symbol of wholeness and unity that was yet to be realized.

## V

## Kept Vows:
## Facing Fears and Breaking my own Silence

I had been keeping a journal since January 28, 1990, and the first entry was a disturbing dream. I felt compelled to document events as they unfolded. When I returned to the United States, my journal had lengthened to hundreds of pages.

Although it was difficult to comprehend, I had been vacillating between shock, denial, and bargaining, stages of grief, for over two years. It was yet another shock when I returned to the U. S., back to the familiar Flatland that stimulated even more memories. I had no place to be, for my office and home were occupied. I stayed with my daughter and for the first time in memory, had no sense of direction. Nor did I have the stamina to be *up and doing*. Adrenalin had kept me going during those challenging years. When the demand for functioning lessened, a collapse was inevitable.

For two months I was a hollow marionette, responding only when my strings were pulled. A permanent string compelled my wooden hand to move across the blank page in front of me. I wrote morning pages, day pages, and night pages of senseless wanderings, sometimes deteriorating into scribble. Those fibrillating lines were a reflection of what was going on, emotionally and physically. That scribble was such a stark contrast to my creative work achieved in Poland, and, at the time, that theoretical work was of no comfort to my sense of devastation. It was as though I had pasted myself together with my thoughts, and I clung to that image of the earth suspended in space.

Gradually, over the period of another month, the earth began to move again. I began to move again in time, a place where, through years of practice, I had gained proficiency in self-reliance and resourcefulness. Both were initiated on the farm and in the community where I was born and my father and his father before him. Throughout my process, even in the darkest times, my timely brain, well developed in pragmatism, performed the fiscal obligations and all other household and personal responsibilities of Flatland, regardless of my inner world. I became increasingly grateful for that ground beneath my feet.

It was my good fortune that, out of the blue, a house was offered to me, a perfect haven where all my seemingly broken parts could find their proper place. It was dry and safe, iron bars covering every window, fireplaces to provide warmth in winter, fans and air conditioners to cool in Florida's summer heat. It was a house that needed a lot of work to make it livable, but I saw it as a project on which I could focus and maintain my equilibrium. The house, in the outskirts of Maitland, Florida, was planted on an acre with large old oak trees and a stream hidden in the back; isolated in the midst of shopping centers, houses, screeching tires. It was where I lived and worked, in virtual solitude, for six years.

I knew I was wounded, one of the symbolic meanings of my wounded golden frog. It was a wounding by the psychologist so traumatic it cost me ten years of my life and irreparable harm to my marriage and my family. It gradually began to sink in that, not only had my life been brought to a halt, almost everything I had worked for was gone—my career, my home, connections with colleagues, most of my friends, and family; everyone except my daughter who also suffered trauma over my catastrophe.

I was also mourning the deaths of my client, father, uncle, and the loss of my companion. Mourning in a way James Agee describes in *Death in the Family*.[37] It was a sensation of heaviness.

I was fifty years old and had little to fall back on except my own inner resources, which amazed even me.

The comfort I so sorely needed came in the form of music. When my compositions began, it was like another voice commenced to speak. Those sounds just emerged. I would write the notes and then have to learn to play them. That musical poetry was an expression of authentic feelings in contrast to *senses learned*. The feedback that music provided gave me strength and moments of joy, as well as consolation. Those flowing sounds also seemed to authenticate what I had seen and where I had been, for like A Square, imprisoned in Flatland, there was no one to validate my experiences.

My collection of poetry, *My Golden Frog* prose, my discoveries while in Poland, and that fourteen pages of words, thick with meaning, were all put away.

I continued my journal and my search for some explanation of all that had happened, including those writings which had also become a mystery to me. I read everything I could find on creativity, and the little I found in all of psychological theory was speculative, at best. I found very little on grieving in the entire field, despite the distinguished work of Swiss psychiatrist, Elisabeth Kubler-Ross, and made accessible in her 1969 book *On Death and Dying*. Grief, though, is not only applicable to death and dying but to the change and loss that is part of living.

The tendency in our "quick-fix" Western society to shy away from the topic of grief has essentially deleted the process from the language of psychology and from the language of our culture. It has become almost as unpopular as the word *victim*. Some, in the field of psychology, like Vann Joines,[38] even contend that there is no such thing as a victim.

Just as *grief* and *victim* are frequently edited out of the counseling and therapy field, Sperry's groundbreaking work on the functions of left and right brain has generally been ignored. To reiterate, psychological theory continues to be a product of socio-economic, political, and cultural factors. Include the personalities and personal environments of the practicing psychologists, and the result is a mish mash of fads often designed for a quick fix and symptom relief. Quick fixes invariably thwart the holistic potential and obstruct the process of development.

Three particular fads I still find intrusive and ludicrous are hypnosis, neuro-linguistic programming, and eye movement desensitization processing, considering that Flatlanders are already living in a trance, programmed, and desensitized. Another similar fad is metaphorical intervention, concealed suggestions, more theorists and therapists playing with people's minds. Whatever happened to the idea of being honest, kind, and respecting the dignity of the other?

What I found most interesting (which could hold the answer to that question), was that in all of my education and subsequent research in psychology, I found little about the human spirit or the soul, even though the first meaning of *psyche* is soul.

The weight of knowledge gained through my spontaneous writings, my visions, creative insights, and that sensation of truly seeing, were heavy enough without validation. The burden was even heavier when I attempted to express, in words, those phenomenal experiences. As I was beginning to discover, the expression of my subjective world had only been in image symbols which have multiple layers of meaning. Those symbols and experiences did not relent easily to definition—"linear terms of dictionaries."

It was sometime in 1994, I do not recall precisely when, I was in a bookstore looking for a book on how to write. I saw, and was drawn to, a small, thin, hard-backed book with a red cover. It was not what I was looking for, but after gleaning through Toni Morrison's acceptance speech for the 1993 Nobel Prize in Literature,[39] I knew it was what I needed. Her powerful, poetic voice pulls no punches regarding the damage language can do and how it is used to thwart our development. In her chosen metaphor for her speech she says that language is like a bird in one's hand over which one has power, language with which one can fly or we can crush it with misuse. Her challenge, to reclaim the power of language, was the challenge I accepted.

I set to work, vaguely aware of where I was going. Finding no help, explanation, or guide that would encompass all the dimensions of my process in the field of psychology, I turned to Plato's metaphor of *The Cave* for some direction and gave it life by adapting it to my own experiences and my own subjective world.[40] (See Appendix I) Although I had read *Flatland*, I was not confident enough to trust its metaphorical portent any more than I was ready to comprehend the magnitude of my visions or creative work. Although there were many of my experiences missing in *The Cave*, there was no question that I needed to move backward in time. As it turned out, I wrote my autobiography and began it by elaborating the symbols of my historical past that were in my Spaceland "box" dream.

Initially it seemed easy to write a linear story, leaving out the baggage, the feelings and events one would rather forget. Out of countless happenings in our lives, those events, significant by the strong emotions tied to them, remain indelible in long-term memory. They are memories that do not go away and are available for recall when attention is drawn to them. That attention may come from outside stimulation, like the wood house in Poland or the child riding behind a tractor. It was my dreams, though, that did, indeed, draw my attention and would not allow me to turn away. I revisited my entire history—the ghosts of my past—with so much emotional intensity, my verbal tenses, in that initial writing, were so mixed up it was difficult to determine if I was in the past or the present.

Throughout my life I had few illusions about the problems in my family. When I married and left home at the age of 19, it was like an escape from a battleground, a civil war between a fundamentalist Christian mother and a secular controlling father, a Linelander if there ever was one. He was master of the home and ruled by fear with just enough brutalizing punishment with a belt or a fist to our mother and us children (an older sister and younger brother), to alleviate any doubt he was boss. He was a man who could also be generous and even kind, but I never knew when or how or why it would change. Nor did I understand my strong attachment to my father.

Dee Graham, in her book, *Loving to Survive, Sexual Terror, Men's Violence, and Women's Lives*,[41] has, since that initial writing, helped me to revisit that attachment from her perspective of the phenomenon of Stockholm syndrome—the bonding of hostages with their captors. Graham hypothesizes that in order for the Syndrome to occur, four conditions co-exist. The first is a perceived threat to survival and the belief that one's captor is willing to carry out that threat. The second is the captive's perception of some small kindness from the captor within a context of terror, and the victim actively searches for those acts of kindness, becoming hyper-vigilant to the abuser's needs, feelings, and perspectives. His or her own needs, other than survival, are discounted. Graham's third condition is isolation from perspectives other than those of the captor. And the fourth is a perceived inability to escape. Graham lists many cognitive distortions necessary to maintain and justify the bond.

My experience of my father fit all of those criteria and certainly contributed to learning my "oh peace" (appease) cry to men.

Even in such abusive conditions my mother did not deny her instinct to bond with her babies. For that I am grateful, for the failure of bonding has dire consequences on development.[42] My earliest visual memory of my mother and me, though, was my mother grasping my hand and wielding a guava bush switch on bare, frail legs, drawing blood, and I was running round and round her, attempting to get away. A good pounding with her fist or a broom, if handy, was not uncommon corporal punishment to correct the "badness" she believed to be inborn, particularly in me. If I reacted with emotion to those punishments, I would be whipped again for crying—or running—which undoubtedly contributed to my "difficult child" reputation.

The warning my mother often used was, "You had better not tell anyone what goes on in this family." Numerous times she said that reputation is everything. Her demand for silence, though, made more sense when I began to realize that my mother did not punish me in front of other people, so there was no one to validate what was going on. The only clue to her subjective world was her assessment that I was just like my father. Her accusation was a source of turmoil, for it was no secret to me that she did not love him. That clue also suggests she displaced her anger from her abuser to me, who happened to have German/Irish facial characteristics, skin, and hair similar to his.

The first draft of my biography was by no means all doom and gloom, for I also had many pleasant childhood memories. In editing my writing I began to realize those enjoyable times were on holidays or Sundays when company would come (always family), and my parents put on their best behavior.

Those days inevitably included delicious country cuisine prepared from my mother's storehouse of home-prepared vegetables, fruit, and meats. Those days involved a tour of the farm, for my father was becoming one of the more successful truck farmers in the area. Then there was, invariably, the story telling where I learned of my family's Flatland roots with a long history of "fog." On occasion there was entertainment when my mother played an old up-right piano inviting voices to chime in. Sometimes her brother would bring his fiddle to play along. Or my father played country songs on his harmonica.

The more I edited my biography, the more I began to detect personal underlying themes and patterns, not only revealing my relationship with my parents, but in engaging scenes that were my means of survival—escape from the uncivilized emotional turmoil in my family.

My first escape was to nature, a source of wonder and beauty, with all its amazing creatures. There were sounds of music everywhere, and I would dance to those imaginary symphonies in my head. I loved nature's powerful forces and all of the seasons. It was and is a love that has never left me, and finding my reflection in the wonders of creation has continually been a resource for insights and renewal. Even my vision of the "I" behind my inside eye, was initiated by a *close encounter* with nature.

Another avenue of escape was the half-mile, bare-foot run through the woods to my aunt's house. I formed a bond with my father's sister even though she did not know it. I loved her. A love built on a single image in memory, so brief, but so real and meaningful: She was washing clothes in the shed outside her house, and I was standing close by and talking. I have no clue what either of us said, just a sensation she actually saw me in a way I had not been seen before—the real me in a very real difficult situation.

That imaginary bond was a major influence in my life. My aunt was a teacher, and I admired all those books on the bookshelf over in the corner just to the left of the fireplace in her old wood house. Those books were a source of curiosity, for the library in my home consisted of the Bible, the Farmer's Almanac, Sunday school picture books, and *The Progressive Farmer* magazine my father subscribed to. I clipped the poetry from the discarded magazines and pasted them in an album. As for reading, there was little time for such activity anyway. My father had a stated belief: when his children became old enough to go to school, we were also old enough to work in the fields in addition to daily chores.

I also bonded with my maternal grandmother, but it was not imaginary. She came to stay for a few weeks at a time, and there would be a period of relative peace. Many times I stood at her elbow and watched her sew on the treadle machine, and she guided my small hands in precise hand stitching and embroidery. She was another model, for as I grew older, I, too, learned to cut patterns and sew on the same treadle machine, the sequential and spatial learning of how to put pieces together into a whole. She made sock dolls for my sister and me, not unlike the sock dolls I made, years later, for my daughter and niece. She allowed me to brush her long gray hair one time which evoked a feeling of intimacy, something so rare in my world. I loved my Granny.

As I wrote my story, it became increasingly a process of discovery with moments of excitement and times when I was simply stunned, like when I encountered the importance of dates and how even they had formed a pattern—the elements of a big picture that moved forward in time.

One pattern, for most of my life, was a recurring depression in January. I had attributed that repetitive altered mood to a letdown after the excitement of the holidays. Then I became apprised of the symptoms that psychologists have rationalized as "seasonal affective disorder." The cause, however, had nothing to do with the season but was an anniversary of another vortex in my life that occurred when I was six years old.

Within a matter of months my aunt moved away,[43] my baby brother was born, and my grandmother died. My mother, lost in her grief and with a newborn, was no longer available to me. At least that is what I told myself for years. There was a scene, imprinted in memory, of me standing on the sidewalk outside the hospital where my mother was with her baby boy. My eyes were searching in the numerous windows for a glimpse of her. I was alone and weeping. It was a time when children under 12 were not allowed to visit in the hospital. Now I know the source of those tears was intuitive; I sensed I had lost my mother. Never again did she acknowledge me as a person, only my measurable accomplishments.

Children grieve.

Grief was overtaking me, and I became gravely ill during the period after my grandmother died. My condition was expressed in a dream, a repetitive dream, as repetitive as the prayer I was trained to repeat, "If I should die before I wake...." In that dream, indelible in my memory:

*I am being chased by a dark tree with no leaves, and I am running from its limbs that are like long arms reaching out to grab me. I run from the tree only to meet a figure made of sand, no face or arms, just moving toward me, and I know it will consume me.*

The running would stop only when the terror wakened me. Even with my eyes open the fear continued, for the same two figures seemed pasted on the wall beside my bed. They were symbols of the terror evoked by my life and the terror evoked by death, the Florida sand in which the dead are buried.

After the effects of the dream subsided, my attention turned to the familiar sounds of the night, sometimes the haunting cry of hoot-owls, the bellows of bull alligators, or an occasional far-off call of a panther, so much like a woman's scream, which would send chills down my spine; most of the time, though, it was the sound of the frogs in the pond nearby that became my focus. If I listened carefully, I could distinguish the different tones—bass, tenor, alto, and soprano—the amazing orchestration of nature, usually begun, it seemed, by a single performer; it was the sound I now recognize as that of a Golden Leopard frog with its upward scaling melody.

That music was a source of comfort that helped me to fall asleep again.

Even though my sister, four years older, slept beside me, I have little recall of her during those early years. She withdrew into her own world of silence, no doubt the safest and wisest place to be.

While music has been and is a consolation for me, I believe the only thing that saved my life, both literally and figuratively, was a new bond I formed with my brother, Charlie, who was born January 27, 1950. It was a sacred bond formed in total innocence, a bond that lasted throughout his life, a bond that continues in my soul even after his suicide on September 11, 1998.[44] That love formed a part of who I am, my *essence*, for lack of a better word.

He was a gentle, sweet child, with fair complexion and the clearest eyes I have ever seen; they were like windows to a soul that is pure. He, too, grew to love nature and became my playmate on that isolated farm. He was also subjected to abusive parents. Even though there was a rise in the family's economic status—a beautiful home with modern facilities—the civil war increased in intensity.

My brother was a victim of a psychological process to which systems theorist and psychiatrist Murray Bowen gives shape—the triangle—when a child becomes the battleground on which parents project their own needs and difficulties.

Charlie was bullied and battered by a father intent on making a man out of him yet gave him everything, a lethal combination. In later years he was his father's "buddy." He had a mother who needed him as a source for her own protection and comfort. Our parents fought over him, through him, and in front of him.

Yet he managed not only to survive but to accomplish some of Flatland's successes. He was popular with girls, star quarterback on his High School football team, and completed 3 ½ years of college. When he went away to school, his emotional dependence became increasingly apparent. His escape, though, throughout even those *successful* years was self-medication with alcohol. A young man of the 60's, that escape extended to include LSD, an injurious contribution from the psychological and psychiatric community, more playing with people's minds.

My brother soon became one of the irregulars, the one who fulfilled the "Identified Patient" role in our family. The Identified Patient is the one who exhibits emotional problems, a response to living with an unhealthy family in denial and the one who becomes the scapegoat and blamed for those problems.

His scapegoat role was not limited to our family but included my father's clan. I recall the ridicule and sarcasm of those bullies toward Charlie; gossiping aunts, uncles, and cousins relished telling "Charlie stories" and laughing at his pain. Scapegoating is a way of denying one's own disowned feelings and projecting it onto others. There were enough skeletons in the closets of that family to fill volumes.

Those scapegoaters also included neighbors, preachers, and essentially the whole community. Small-town policemen, who, more often than not, hiding their own hostilities behind their guns and badges, would pick him up for anything, including loitering, and cart him off to jail. He became the brunt of jokes and jeering, gossip and isolation, the group behavior that kicks the one who fell down. Of course he wore psychological labels that describe the symptoms, not the disease, of a complacent, hypocritical society that degrades human worth and condemns its victims. It takes a village to raise a child.

Yet in the writings I discovered after his death, I can only surmise he remained a poet at heart. I read one of his essays, which speaks of the brother I love, as a eulogy at his funeral (See Appendix II). I was also surprised to find in one of his journals he wrote at a drug treatment center, descriptions of synchronicity—encounters with nature and the Spirit of life—for which his Christian counselor chastised him, in writing, in Charlie's journal. It was only one of many "treatments" he received for the symptoms, not the cause. He would leave those in-patient facilities "clean and sober" with a demeanor that screamed to me *contrite and powerless*, bereft of the resources or support to sustain a sober state.

The drug/alcohol treatment centers with which I am now familiar are no different from the system that oppressed and abused him, for they include the tactics of intimidation, scapegoating, and bullying in their therapies. More closed systems. I say that from experience as a counselor at one of those centers. (Described in Appendix III)

It was so poignantly symbolic that my brother's final escape was made the day he freed a neighbor's goat that was being fattened for slaughter. He was too gentle to survive the destructive energies of Flatland. Even in those early years I defended him and tried to protect him but was helpless to save him.

My love for my brother expanded through the years, and I acquired the language to defend anyone who was oppressed or maltreated. In my youth that defense included the black women I worked with in the fields on a daily basis. They became my friends, women who would actually talk to me and listen to me without judgment. There were scenes in my memory when I took a stand to defend my sister, the Jews, and all people of a different ethnic background, then would pick myself up off the floor after a fist-in-the-mouth beating by my father.

Love for my brother, in later years, helped me to understand, empathize with, and feel true compassion for my clients in the therapy room, something I knew but did not know I knew until January 27, 1990, Charlie's birthday. I was walking beside a young male client who was pacing in his distress. Suddenly my brother's face appeared in the face of my client's! It was a jolting experience that provoked my strange and disturbing dream, a dream I did not fully understand for many years.

Both my mother and father, though seemingly opposite, were the same in executing the oppression of a closed system. My mother sabotaged my friendships by unrelenting criticisms of friends and their families, rarely allowing me to visit or have friends over. My father also sabotaged every effort I made for a normal social life, like threatening young men who were interested in me. Or the time I had the opportunity to go skating with the members of the Methodist Youth Fellowship. My father forced me to work in the field until almost time to meet my ride. I heard his scoffing laughter as I struck out on the mile-long, hilly road to the house, driving as fast as the Farmall cub tractor would go. I had half-an-hour to scrub the field dirt from my hair and body, then dress. I made it.

I was also keenly aware of my *house arrest* effected by my father's threat, "If you ever try to run away, I'll hunt you down wherever you go." I believed him. I was also aware of my mother's rigidity. While I could see the cruelties inflicted on my brother and sister, I could not see another real and lasting harm inflicted on me, for it was not visible. It was not just my mother's abandonment or the "punishments" by my father and mother that damaged me. My brain had been burned by my mothers tongue and the tongues of fundamentalist preachers that incite terror in one's mind of hellfire and damnation, apocalyptic teachings, the rapture with "non-believers" being left behind; words that "drink blood and lap vulnerabilities."[45]

When I was a teenager, it was the one-liner responses if I balked at my mother's Sunday school teacher hypocrisy: "You're going to burn in hell. God knows what you are thinking and thinking sin is the same as doing it." Her Victorian tongue shamed my human needs for closeness, even human touch, my femininity, and my 5' 2" body which, ironically, was shaped like hers. Her terror inciting repertoire extended to, "You're crazy, I'll have you sent to Chatahoochee," a state mental institution, when I dared to mention a word remotely similar to *evolution*, or questioned Christian beliefs.

During those teenage years I made attempts to write about another plane of existence—holistic vision—and burned those stacks of paper for fear that my mother was right; I must be crazy. My writing also included a thread of consciousness into which one could tap. It was an attempt to explain premonitions for which I had no other explanation. The feeling of insanity, though, won out. It seems the inner turmoil was resolved with another escape as revealed in a repetitive dream recorded in a little notebook:

*I was climbing a rickety staircase, the steps sometimes breaking beneath my feet. I could feel the terror as I grasped onto the railing to keep from falling. But I always climbed on to reach a level that contained the mysterious part of the house. In its day it was of fine quality, but neglect and time had reduced it to a hopeless state. It contained many mysterious rooms with furnishings in various stages of decay.*

*In that dream I was thinking that perhaps someday I would be able to use some of those rooms when I had the resources to repair them. Then, with a feeling of despair, I walked out of the door.*

That other hemisphere, my spatial brain, developed separately but paralleled my life. I have recollection of only three dreams during the ensuing 25 years. From that point on my biography was, indeed, a story of struggle with the exception of the birth of my child who was the light of my life.

Perhaps it was just one of those *random acts of kindness* when, half century ago, a Methodist minister saw a lost teenager. He offered no advice, just a helping hand and a book of poetry in which I found Longfellow's poem. Now I suspect he was intuitive. He, too, was essentially a mute and was banished by the members of the Linden

Methodist church because he lacked the skill of rhetorical preaching. It was another loss for me of the validation I so sorely needed.

༄

As time progressed on my special acre, my writing developed into another kind of reflection, on what was referred to in *My Golden Frog*, as "my own debris." Self-reflection was also foretold and directed in one of my poems written in 1990.

> Spirit requires a disciplined mind
> with faithfulness of purpose
> for revelation to be complete.
>
> To deviate, mind does forsake
> and calculates to seek,
> with ease, another story.
>
> It looks in mother's face
> to find flaws it would hide;
> its own mirror points the finger.
>
> Past is dead I would believe,
> now it's time truth to seek,
> mistakenly without discomfort.

I had intended to provide a better life for my daughter and me. I was determined to protect her, and nothing could rile me more than a threat to her safety. Yet I had exposed her to the violence in my family. She was also exposed to a mother who was forced to tolerate oppression and ridicule—more insults and injuries—by her father and my first husband, a reasonably successful Flatlander.

My first attempt to leave him was thwarted by my father's shotgun wielding intimidation to a friend who tried to help me and my mother's terrorizing insanity threats after kidnapping my daughter. I knew I had to yield or something tragic was inevitable.

Then I became a mother who frequently experienced bouts of depression and physical illness. The only means I had of combating those depressive episodes were projects on which to focus. I learned bookkeeping as a skill, studied art, improved my music, and even progressed to embrace the less oppressive Episcopal Church.

The dream I experienced just before finally leaving him was exactly like my repetitive childhood dream: *I was being chased by a tree with no leaves, its limbs reaching out to grab me.* The odd part about that dream was the image made of sand was missing. Death was no longer a threat to me; my fear was one of living. I drew a picture of my dream, and visually exposing that fear helped to temper it. I mustered the strength to face the condemnation of my family, friends, and the whole community when I moved out with my daughter to go it alone with meager means.

Two years later, at the age of 32, I had the opportunity to leave rural Sumter County, a major change in my life. I had been offered a bookkeeping job at the Episcopal Diocese of Central Florida and moved to the city of Winter Park.

Then I became a mother who, for all intents and purposes, abandoned her daughter to become super-woman, climbing the ladder to success. That (mother) was who I used to be, a person I am no more. Nor can I identify with the half-brainer I was when I was proud to be a self-made woman, for beneath that façade was all the emotional baggage that showed up in my relationships, especially my roller coaster rides with men.

In April 1984, nine years after I left Sumter County, I met my companion—another life-altering event—when I was just coming out of my second marriage to a banker. The song, *Muse*, I have written about that love best describes our relationship. The main theme or melody is a chromatic scale that touches every key. He awakened me to sensate existence, for it was in our relationship I experienced what it was really like to be a woman. We were a perfect match, as though our bodies and unconscious minds knew something we did not know. There were remarkable instances when we sensed each other's feelings even when we were apart.

In the beginning I did not know that he had also been damaged. We would often reach an impasse when he would withdraw into his world of silence. His withdrawal was a fiercely disturbing experience for me, what one might call *abandonment issues*. Other times he would run away, but he always returned. I loved him, a love beyond rational. That awakening to sensate existence aroused awareness of my spatial brain, for after we met, I began to recall my dreams again.

In 1986 I had an extraordinary dream recorded in a journal. It was the same dream I had as a teenager, but different, for there was no fear-provoking staircase to climb:

*I opened a door and there it was—the spacious house, still old and dusty, still of fine quality but not irreparable damage. A good cleaning and polishing would take it back to its original state. In the new dream I began making plans to live there. In the new dream I did not leave, and the dream ended with hopeful, good feelings, thoughts of healing and repair.*

## VI

## Psychology and Religion without Soul: Discovering my own

I had been fascinated with dreams and dreaming most of my life, which sparked my interest in psychology. My entry into the field, though, was gradual. Although I had not given up my interest in the natural sciences, my goals had changed through the years to something more practical and in keeping with a variety of experiences in the work field.[46] When I began college at the age of 32, my plan was to achieve a business and law degree (even though my first course at Rollins College, Winter Park, was painting). Introduction to Psychology was interesting, but I became committed during my first Abnormal Psychology course.

A case in the textbook especially impressed me. It was the story of two sisters labeled with *schizophrenia*. The report of the horrendous torture of those children by fundamentalist religious parents, including mutilating surgical removal of their clitorises, confirmed something I had come to believe—the relationship between violence, both psychological and physical, and so-called mental illness. I completed my Masters Degree in Clinical Psychology at the age of 42.

When I actually began to practice in the field, I believed in its code of ethics that speaks of human values. It was a gradual awakening over a period of years, to the lack of human values in practice. Initially I began to question the psychological testing, including IQ. The norm is based on Flatland measurements and weighted against the irregulars. I began to question its major book of reference, *The Diagnostic and Statistical Manual of Mental Disorders*;[47] it was in direct opposition to those codes as were testing practices.

Labeling a person is dehumanizing to the person labeled as well as to the labeler. It narrows one's perception of the other, and it becomes practically impossible to actually see the whole person. Teaching those labels is more indoctrination (like Abbot's A-Square, who acquired through his education a keen sense of hearing and seeing), the mental fog altering sensate existence. It has taken two decades to override those labels in my mind.

As I advanced in my practice, I refused to use any diagnosis other than one of adjustment accompanied by the stress involved; the extreme being the stress reaction to inhumane trauma. The only other diagnosis I would use at this point would be on a spectrum of developmental impairments—the failure to mature or evolve consciousness. The failure to mature would include the many therapists I encountered throughout my career who were also carrying their baggage of disowned feelings.

Before entering the field as a practitioner, I was already alerted to the possibility of peril in the therapeutic process. The "Big Daddy" of psychiatry in Winter Park and the official psychiatrist for the Episcopal Diocese of Central Florida had advised me to stay in my marriage to the banker and entertain myself with extramarital affairs. I later learned he had a personal vendetta with my husband.

A part of my education was when I noticed some of my fellow students, professors, and college counselors were enmeshed with religion. An experience with one of those counselors was eye opening and enormously disturbing when he judged me through his religious beliefs. Just as disturbing have been so-called professionals who use Continuing Education courses to indoctrinate more therapists in both new-age and old-age dogmatic closed system thinking. Psychology as a *science* became increasingly ludicrous.

Another personal experience, very early in my career, was during a job interview with a psychologist, a specialist in psychological testing, who exhibited the rigidity of a true Flatlander. All of his language was demeaning toward me, and my intuition told me to run. I was not surprised but deeply saddened when, years later, his teenage son committed suicide.

The biggest shock though, came from a psychoanalytic "Self" psychologist and supervisor, who I perceived to be a very effective and capable therapist. It was in social conversation that he defended priests who have sex with parishioners. His defense was that women are seductive like his client who had seduced him. When I confronted him with the fact that sex with clients is unquestionably destructive, his retort was that I had a small-town mentality.

It was not only disowned feelings, personal agendas, unkindness, or therapists without personal boundaries I discovered. There was also corruption. I recall telling a friend and colleague I once admired, of a psychologist who was exploiting Medicare in his nursing home testing practices. I was taken aback when she said he simply had a different view of the world, and I must learn tolerance.

As I have since discovered, the mental health care system, like our systems at large, protects those who would corrupt anything and everything that is good to the point of altering the minds of our most vulnerable, our children and our elderly population, with injurious drugs.[48] Perhaps by being in bed with the drug companies, through their generous financial support, the American Psychiatric Association and American Medical Association have obstructed valid research regarding the poisoning of our children with inoculations that may thwart the development of their brains.[49]

I found my niche, though, in private practice. My reputation as a therapist who could work with *difficult patients* began to grow, as did my caseload. I did not see my clients as other than equals, symbolized by two identical chairs in my therapy room. While my work, at the time, seemed to be primarily intuitive, I now know I understood their language of symbols, expressions of the unconscious. Those symbols, when acknowledged and attended, serve to evolve consciousness. I also saw their symptoms as symbols representing the complexities of survival, including cognitive distortions, erroneous beliefs about themselves, and the baggage they all carried. I was constantly astounded at the atrocities visited on both children and adults.

The source of one of my poems was a client who evidenced the symptom of bulimia. Over time she revealed the familial savagery experienced throughout her childhood and youth.

> Despair I attempted to evade,
> the word that suicide is made of,
> the facing of the truth of evil,
> no chance of reconciliation.
>
> Children bear the horror,
> the damage goes unseen;
> silent, killing evil does exist
> in the eyes of complacency.
>
> Turn the other cheek from the vomit,
> the child cannot stomach the horror;
> call it mental illness, label it if you please,
> death of this society.

Increasingly, my initial thought about the relationship between abuse and so-called mental illness was confirmed, long before I became aware of the work of Alice Miller, renowned European psychologist, whose insights regarding mental illness were similar to mine.[50] It would be some time, though, before I could comprehend the full portent of what my poem was saying. Especially disturbing was the word *evil*, for it had many religious connotations for me. I had essentially deleted it from my vocabulary.

The young subject of that poem was also instrumental in my introduction of art into sessions when she brought pictures of suffering children to share with me. They were symbolic of those things she knew but could not or had not been able to verbalize. Then there was the inevitable grief when she began to confront her memories. Her capacity to heal, reclaim her life, and move toward maturity was remarkable.

Another observation evolved into a belief that a "wall of words" can be just as difficult as a wall of silence, for both are defenses against authentic feelings and the things one knows but are too fearful to face. Art, even the most rudimentary drawings, became a means of breaking through those walls. Unfortunately, it was rather late in my career when that lesson began to sink in, even though early on I had used a genogram—a pictorial display of familial history and relationships—to identify repetitive patterns of behavior.

I truly valued my work as a therapist. While writing my personal story it became increasingly valuable, for I began to realize that it was my clients who provided my true education in psychology.

*Disparate* is an appropriate word to describe my life before my awakening. From a Flatland perspective, it seemed that I had all one could wish for. Along with my professional success I had a beautiful home, an elegant lifestyle, and friends from the art community with whom my Polish architect lover-turned-husband and I mingled. Something, though, was missing. Sometimes I felt as though I was a great pretender, often alone in a crowd.

The conflict was exposed in a dream. I was unable to reconcile the suffering world evidenced in the therapy room with my *comfortable* life. Some of my writings in a little diary also suggested that comfortable life was a deterrent to my true calling—my real purpose—which, at the time, I had no clue.

༺

During the initial writing of my biography, I was continuing my journal where I toyed with theories of what had happened to me with the psychologist. Especially how I came to be so diminished I wound up in a psychiatric unit for four days, an attempt by my so-called therapist to force me to recall "repressed" memories of sexual abuse that did not happen. Had it not been for the intervention of a nurse, who knew me as a therapist, and a social worker, the nightmare could have been worse. The social worker witnessed the psychologist's terror-inciting, threatening language and confronted him. I believe that I will always remember her words, for it was one of the few times in my life that anyone defended me, "You can't Baker Act[51] Frances because she isn't crazy." They were words that broke the bonds of enmeshment.

The clues to how I became so entrapped were in my journal, which was too painful to read at the time, and a poem:

Disembodied memories
paint surrealistic pictures

to disturb my soul,
yearning to escape
unspoken, broken pieces.

Specter from hell,
taunt me no more
with bewitching
intentions of
unspoken, broken pieces.

Attend your own
pale countenance
seeking to dissect
me to find your
elusive soul.

My journal began to reflect another gradually evolving theme. For a long time I believed my solitude was voluntary, but it was not. I had become an irregular like my brother. Every attempt to reconnect, even with former colleagues, was met with platitudes that in time diminished to silence. It is a silence that becomes the prison to which many irregulars are confined, with walls there is no way around or over or under. The resulting loneliness, at times, seemed unbearable. I no longer displayed the trappings of success in Flatland.

Dee Graham[52] uses the term *defensive attribution*—the tendency to blame the victim—to describe my isolation. It is a distortion or distancing that helps to defend against the idea that *no one* is impervious to the destructive events that do occur. Graham also suggests such distancing is especially significant when the victim is socially respectable.

The hardship was enhanced when attempts to re-establish myself professionally failed. Eventually I accepted any menial job that did not challenge my integrity. A dimension of that integrity was to accept only part-time work and make whatever financial sacrifices necessary to buy the time to keep my vows.

# Psychology and Religion without Soul: Discovering my own

In early spring 1994, after nine months on my special acre, my journal writings progressed to the idea of closed systems and my own dependence—my need to be seen, heard, acknowledged. The word in psychology is *enmeshment*, one's mental entanglement with another. While it is similar to Dee Graham's ideas of Stockholm syndrome, which I discovered a decade later, I contend a broader perspective and a more subtle means by which one becomes ensnared.

Enmeshment, too, is the result of a process and is the means by which people become trapped in their family of origin, trapped in relationships, trapped in sects, trapped in oppressive ideological systems, trapped in religion, trapped in the religion of psychology, trapped in any other closed system.

Need is a prerequisite for vulnerability to enmeshment. That need may be a basic biological necessity of food or shelter; it may be an esteem need, social need, need for a sense of purpose, or any other of an astonishing array of human needs.

Another part of the enmeshment process is the promise of need fulfillment to maintain focus. It is a process that plays not only on the needs but also on the fears of people, the forming of double binds. Secrecy also plays a major role to conceal the motivation or agenda of the person in the position of power. It is a process that thwarts open communication and any dissenting voice.

The most effective scheme, though, is isolation. That isolation is effected by judgments and criticism, both subtle and blatant, of those who disagree. It is the means by which beliefs about oneself and others are taught; the means by which prejudice is instilled in one's mind: the scapegoating of a person or group of people. It is the persuasive brainwashing by which the person in the position of authority—whether it be a parent, therapist, president, priest, preacher, guru, pandit, Hitler, or Osama bin Ladin—shuts the door and turns a potentially open system into one that is closed. Some systems are more destructive than others but the process is the same.

Necessary to the process of enmeshment is the capacity to distort the world to fit those beliefs, the compulsion to defend them, and the capacity to shut out any conflicting evidence.

Simultaneously, while discovering, organizing, working on, intellectualizing that process, and recording it in my journal—linear brain activity—I was also deeply involved in editing my story where the mental entanglements were becoming increasingly obvious. Then one day, while engrossed in my work, my mother telephoned, and we began to argue. As I recall, the subject was Charlie, my brother. Her familiar words, "God knows what you are thinking, and so do I," evoked a tightening in my body, and my unspoken reply was, "If you really know what I am thinking, I doubt you would be talking to me."

After that event, attempts to write proved useless. I would get up from my desk and pace then set in to clean the house, a symbolic behavior on my part I had come to interpret as a need to clean up my emotional mess. During the ensuing week my behavior became progressively fanatical, alternating between pacing and cleaning. There was no question that it was obsession when I woke from a stupor to find myself on my knees behind the door of the utility room scraping at a tiny spot of white paint on the vinyl-tiled floor with a needle. Yet I was unable to stop. I am uncertain of the sequence or even the number of days my mind was so immobilized.

Then on the morning of April 14, 1994, I opened the back door, and a dead cardinal was on the doorstep. Suddenly I heard the words and melody of a song my mother used to love to hear me sing, "His eye is on the sparrow, and I know he watches me." The words just seemed to come out of my mouth as I looked at that dead bird, "There was no one watching out for you."

Without warning something snapped, and in an instant I made the connection between my mother's projections—her own view of the world assigned to me—and the projections by my therapist and countless others in my life. It was also a sudden insight into the meaning of the hideous mask I drew upon entering Spaceland. It was more than a peak experience, for my linear brain, running at high speed with flashes of memory, met with my spatial brain, and once again, I could see everything.

The seeing experienced in my initial awakening was about the world outside of me. This seeing was about the inside of me. It was a sudden insight into what those empty eyes of the mask I drew could not see. It was an image of insidious mind control meant to silence every thought conflicting with patriarchal Flatland and every feeling that did not meet with approval.

Although *My Golden Frog* writing predicted grief, including rage, it did not prepare me for the pain of days as they turned into weeks of screams that wracked my body and my mind. I had encountered, in that instant, the cost of my enmeshment in the closed systems of religion and the religion of psychology, the cost of my dependence and idolizing men.

It was also a sudden insight into a marital therapy session turned nightmare. The psychologist is a Hannibal Lecter[53] who, like fundamentalist preachers, drinks blood and laps vulnerabilities, playing on my human needs and fears, pulling me into his story, then leading and trapping me in his net of words. He is still a practicing psychologist and thousands more like him (a few of whom have been prosecuted), who continue to toy with the human mind with no remorse.

I was one of the more fortunate ones to survive the false recovered memory debacle. Countless women are still lost and many committed suicide. There was a brief period, after that initial wave of rage subsided, when that "escape" was very enticing, for I finally faced the truth of evil. My sense of devastation was so intense it seemed I could not endure.

I had a vivid dream portraying what I was feeling:

*It was a black slave I had previously freed from chains around his hands and feet and had removed the blindfold over his eyes; he had been attempting to climb a steep hill on his knees. I told him it would be easier to climb without his blinders on. Then the slave became a man bared of flesh, raw and oozing. An image suggesting he had been tarred and feathered.*

"Tarred and feathered" was one of the terms used in the stories told by my father in reminiscing on the culture of the Rebel South. It was a vigilante punishment to slaves who stepped out of line, meant to humiliate and scar for life.

It was truly as if my entire self was an open wound. Sounds became unbearable. Even the music that had been so much a part of my life brought agitation. Shopping for necessities became a dreaded challenge; sights, sounds, and movements seemed to penetrate my brain like flashing strobe lights. I could leave the safety of that acre for brief periods, no more than a few hours, by talking myself through the drive or the tasks required for survival.

I began to see the truth of my life. I had not been in control of anything. The human waste piled in chaotic heaps became increasingly personal as I realized what had happened to my brother had also happened to me. My obsessions had simply been more socially acceptable. I began to experience the humiliation my brother, no doubt, endured and sensed his condition was terminal.

It all became clear one day as I was working in my garden. Out of nowhere, images of Charlie began to flash in my mind, and none of them were pretty. They dissipated any denial regarding his destruction. I began to walk, to weep, and felt the urge to run. It seemed as though all the broken pieces of me were moving chaotically. I began to hear again the eerie sounds I heard at Majdanek, the death camp in Poland, and a similar chill surged through my body. I raised my arms, shook my fists at the sky, and the voice screaming in my head came out in a whisper, "Remember, he is my brother!"

In the space of an instant I heard an entire orchestrated piece that took four hours to sketch and another year to complete, for I lacked the knowledge of how to layer all of those sounds. It is a piece of music that has been described as a canon and its shape that of a palindrome, the same backward and forward. My song, *Reflection on a Prayer*, was indeed a reflection—of a new woman of many dimensions with a center that is immeasurable and ill defined by humans. My love for my brother had saved me once again.

Leonardo da Vinci believed the location of the soul is between our dual hemispheres.[54] In a sense, it is. Now I know it is the "I" behind our insight—the soul—that is the unifying center of all aspects of the self, the space-time-matter-energy continuum.

My soul provided the support I needed to go through that frightening, shadowy doorway in my "box" dream. It was a veil of tears, more like a shroud of death. I was comforted, though, by the memory of the *correcting* voice I heard six years before, "Don't take this so seriously, it has always been." It has always been—grief is a part of living that brings closure to the past.

I put everything away but my journal. I needed to heal and to process all of my discoveries. I wept my river of tears for the countless insults and injuries to my body and my mind. And the question that confronted me was, "Where do I go from here?" For another year I was a wanderer, adjusting by degrees, as though learning to walk again.

In early 1995 I ventured out of my solitude to attend meetings at the "The School of Wisdom." I had heard its founder speak at a seminar in 1992 before moving to Poland. From its leaders I learned a lot about energy flow, brain waves, and stages of consciousness, especially during dreaming and creating. I learned how music and sounds affect the entire body, especially the brain. Music can also play a major role in both learning and memory. It was also a confirmation of something I had long suspected. Even music can be used to seduce one into altered states, like patriotic marches or religious hymns. Perhaps the reason I needed to write my own.

They delved into chakras, numerology, and the symbolic meaning of each color. It was the epitome of a school engaged in *mystic, twinkling obsession*. There was absolutely nothing about the baggage we carry from our past, nothing about grief, and nothing about modesty. That was evident when I was re-introduced to its founder (a colleague of Carl Jung), whose self-omnipotence was obvious, as was the *omnividence* of his followers.

I went along for a while until the leader began training in rituals, evoking a knee jerk reaction. I knew from experience what rituals do, those religious beliefs I had been trained to repeat; I contend they form pathways in the brain, connections one cannot undo, only initiate the development of new ones by expanding one's knowledge.[55]

Since then I have thought a lot about the Flatland of closed systems where the holistic spatial hemisphere is not acknowledged and not used. Unfortunately, it is not that Flatlanders WILL not see the big picture, but many cannot (developmentally impaired), unless or until they engage in a dynamic process of personal growth. To quote Betty Edwards, "Half a brain is better than none: A whole brain would be better."

Those new-age Spacelanders, although apprised of the creative capacity, seem to be missing another essential part: the self-knowledge that comes from examining one's own history, which often becomes a process of self-discovery.

"An unexamined life is not worth living" was Greek philosopher Socrates' famous statement. It is not clear if he was referring to one's own history, but certainly the freedom to ask questions.[56] There were no openings to question the *wisdom* of the school, which suggests it was not so wise. Closed thinking, be it new age or old age, has the same result. It is not either creativity or rational, not spatial or timely existence, but a synthesis—the balance—of both that brings us to wholeness.

I knew then the meaning of the phrase, "mystic, twinkling obsession," in *My Golden Frog*; it is but another avoidance of the grief involved in closing the doors to the past.

I was fascinated, though, with a book introduced at one of the "School of Wisdom" meetings, *The I Ching or Book of Changes*.[57] In the introduction, Carl Jung asserts the basis for its use is that it works through the unconscious and through synchronicity. My fascination was a philosophy based on the wisdom of nature. Also, its form, presented visually and in poetic, metaphorical language, was engaging to my spatial hemispheric inclination. In many ways, it became my teacher. Every appeal to its wisdom regarding my writing produced the same metaphor encouraging me to continue.[58]

I retrieved my biography and found a woman to help me edit and expand it. She corrected my language and gave me the feedback I needed when she found holes in my story, my blind spots, sometimes indicative of the things I did not want to address. With her support I became increasingly honest with myself, essential to movement toward maturity.

I also found a teacher to help me learn to layer all of those sounds in my head—my musical poetry—that continued to flow. In the 2½ years I studied with him, my composition teacher taught much more than music. He was one of the few men or women I had ever known who was truly and consistently kind to me. My music flourished in that environment which was a stark contrast to my initial education in music, essentially learned on my own.[59]

*A Mother's Song* expressed my love for my daughter. *Turning the Stone* was evoked by my grief over the death and life of my father. The story produced by the form made me realize he was not all bad or good but the product of his family and the time in which he lived.

In editing my writing, I became increasingly aware the realization about my father was predicted by my "box" dream preceding my encounter with Spaceland. While my biography addressed the various symbols of my history exposed in dream image, I began to interpret the meaning of the dream holistically. Those generations before me, including my father and mother, were as trapped as I was by the patriarchal beliefs of Flatland. It was an understanding that moved me to acceptance, the final stage of grief.[60]

I also began to fully comprehend that my teenage house dreams, which culminated in my Spaceland "box" dream, were not natural but about the "me" constructed by Flatland. It was an understanding that evolved from my major insight in Poland—the flowing form of nature. It made me realize the extent of the power patriarchy wields by constructing prisons in one's mind.

As to my own debris, I knew I could do nothing to change the mistakes of my past except to learn from them and vow to change.

In February 1998 I began to paint again. I had given it up many years before when the contents of my paintings frightened me. They were symbols from my spatial brain telling me things I knew but did not know I knew. That February painting, *Primary Colors*, was so very different, for it was evoked from a place of joy, a celebration over the birth of my beloved grandson.

After my brother's death, seven months later, both my music and art supported me through another period of pain, a period that seemed as though it would never end. By then, though, I had learned to make friends with time.

It would be another two years before I could integrate the song I wrote, "A Winter's Night," and the painting, "Funeral." They are inseparable in my mind, for both are reminders of a wondrous, mystical scene encountered in Poland.

My companion and I were traveling from Warsaw to Bialystok late on a moonless night. Only the headlights of the car pierced the darkness. Suddenly we came to a clearing flooded with light as though energy was emanating from the earth, an energy that cast an ethereal glow on the snow-covered trees and ground. That scene, so chilling and mysterious, is joined in my music by the haunting sound of an owl perched in the huge oak tree on the rented acre in Maitland. Interspersed in the song are the pleading sounds of the canon written previously, "Reflections on a Prayer." There are also the tonal words, *the bitter and the sweet*, the movement between minor and major keys, and the ending, a lullaby, "Now I lay you down to sleep..."

My grief over Charlie's life and death came to resolution in a series of paintings in a single afternoon that turned out to be the faces of grief: shock; denial/bargaining; anger; despair; and acceptance, and in the words of a poem. It is a poem that speaks of the sacred—my undying love for my brother:

> Young goat running
> down the canopied road,
> freed by the villains hand,
> villain had run that road before
> a thousand times still bound.
>
> Not once did a man,
> stumbling or lying in a ditch,
> fail to jerk my eyes, stop my
> heart and breath,
> perhaps it might be.
>
> That villain whose eyes
> and heart and breath
> I shared, two children
> walking hand in hand
> under that same canopy.

> Bitter and sweet the memories
> of the hand that tied
> the noose round my
> brother's neck,
> the goat, at last, is free.

In the interim, in July 1999, I left my solitude and made a major move to Tallahassee to work on my relationship with my daughter and to put into practice all I had learned, especially using my *paper* voice, in real time.

## VII

## Real Time:
## Evolution out of the Flatland of Closed Systems

It took *real* time to appreciate the difficulty of maintaining a sense of balance and stability in Flatland. I began to understand how re-entering Plato's cave, back into the darkness, one stumbles and falls. Returning to the Spaceland metaphor, any major change requires adjusting by degrees (the process of adaptation).

Even though I had learned, during my six years of solitude, to be frugal, the continuous task of acquiring the basic needs of food, shelter, and clothing was arduous. In spite of working part-time at whatever job I could find, all of my resources from the sale of my home and office had been depleted. I found an old fixer-upper house and another part-time job as a bookkeeper; just enough to make ends meet. Working on that house was another project I needed in order to "see" achievement (short-term reinforcement), while meeting the long-term challenge of overcoming my history with my daughter. I was not ready to face another phase of confronting my past.

At the root of that lack of preparedness was a blind spot to the influence of a so-called friend. He was skilled with words, mathematics, science, and entrenched in psychological theory. His theme only became apparent over time: an arm-chair pseudo-counselor with a load of baggage from his past. While he appreciated and complimented my art and music, he discounted all of my creative thought. At first it was my poetry, then *My Golden Frog*, then my idea of the space-time-matter-energy continuum as it relates to human functioning. As previously described, in the last days of 1999 I wrote the chronicle of *Flatland* and with a no-confidence vote from him, I also put it away.

In January 2000 I did a major painting titled *Let It Be*. It is a picture of a woman's face that paints more than 1000 words, for it speaks of the many dimensions—the thickness—that can exist in a single image. Now I know the large, pensive eyes reflect my visual-spatial learning preference. The mouth is closed and silent, symbolic of the inadequacy of language "in conscious conversations" to express one's whole self. The right ear seems open and prominent. While learning to use language had become important to me, it was time to hone another skill—listening for the damaging words so carelessly used, including my own. I knew that image was also the visual rendition of the Beetle's song by that same name, for I had played it incessantly in Poland just to hear those few consoling words, "Let it be; let it be. There will be an answer; let it be."

Those words were applicable to my relationship with my daughter, applicable to my own gradual development and evolving insight. A part of that insight was that untoward influence of my so-called friend. It would be another year, as I gained distance from him, before I began to write again.

That writing was to face the most horrendous event of my life: the destructive encounter with "Hannibal Lecter." It was a subject I approached with dread, and now I understand why. One cannot look at evil without being affected by it no matter how many times one encounters it. It is not something to which one can or should ever be able to reconcile or make normal.

The details recorded in my journal throughout the 18 months I was in his power revealed how he progressively and systematically devalued many of my thoughts, feelings, and all of my unusual experiences; interspersed, however, with compliments and just enough mirroring to make me believe I was being heard.

It revealed how he discounted my poetry and my art when I attempted to *show* him what was going on inside of me.

It revealed how he criticized every person in my life, including my friends, colleagues, family, husband, and daughter.

It revealed how he sabotaged my marriage and even stated my husband was attempting to drive me insane, then his support of a divorce.

I can only surmise his intent was to isolate and gain control over me, for that is what he accomplished as he progressively, through the power of suggestion and the trickery of paradox—a double-bind mind game—convinced me I had repressed memories of incest with my father. With extraordinary and accurate visual memory, I was not ever able to even conjure up such an event.

My *appease* cry became alarmingly apparent as I read my journal, and the conflict between my so-called conscious thoughts and those things I knew, unconsciously, was blatant. His words that I trusted and came to believe were frequently recorded on the same page with dreams that refuted those words. My journal revealed that I had the same response as I did as a teenager—there must be something wrong with me.

The false recovered memory debacle has been described as a cult-like entrapment, and in *Remembering Trauma,* as a "disgraceful therapeutic craze."[61] Another aspect of the disgrace are the countless real abuse cases that have, through many years, been ignored. There is no evidence, scientific or otherwise, to substantiate claims of repressed memories of traumatic events. Nor is that issue controversial to me. That fad is the result of pure myth, and one that is nothing less than evil in the untold damage it has done to countless lives in the witch-hunt it caused. It was a witch-hunt comparable to the dark ages. To my knowledge, few have explored the obvious question: How could such a thing happen? I have only heard tidbits such as "bad faith" practices.

The book I wrote about that despicable experience was a process of discovery, for it was in that writing, I found my own answer: the system of psychiatry and psychology is another closed system. The concept of "closed systems," though, was practically missing in regard to our human systems—unexplored territory—with the exception of a few family theorists in psychology. Using their words, I patched together a definition:

> It is a system that exhibits consciousness in its own unique language (in its labeling and divisions of the self) and a system with a "collective unconscious"

> (cultural bias) of patriarchy. Control and power, in the name of authority, are basic tenants, so much so that new information is relegated into an ever-increasing number of subsystems.
>
> Just as any other belief-based system, such as Christianity, has split into diverse denominations and sects, so has psychology. The major theoretical models have also splintered into numerous fads, some of which are so mind intrusive, they qualify as criminal: hypnosis, multiple personalities, recovered memories, past life regression, neuro-linguistic programming, eye movement desensitization processing, rebirthing, the child within, forgiveness, gurus, inspirational speakers, coaching ... the list goes on. Reforming or transforming the system as a whole is prevented by the system itself—the self-governing, belief based entity—a closed system. The potential for harm by omission or commission is intrinsic in a closed system.

While one could assume the variety of concepts within the system of psychology is reflective of an open system, it is not. Instead it is an indication of psychology's deterioration toward entropy—a measure of how disorganized a system is—which is also a reflection of our human systems at large. It goes without saying that another defining principle is that a closed system lacks transparency that would expose the malevolent alliances and practices within the system.

One of the most revealing books I found regarding that system is by Jerome Frank, Ph.D., M.D.[62] who suggests that the theories—the product of subjective confessions—actually do not seem to matter in the therapeutic process. His 20 year endeavor to determine the outcome across a broad spectrum of theoretical models resulted in his conclusion: "It is not the mastery of technique or the belief system of a therapist that accounts for positive change, but rather his or her personal qualities" (the person). He also asserts many therapies are surprisingly similar to rhetoric (persuasion) and hermeneutics (interpretation).

*Persuasion and Healing*, the title of his book, provides a clue to the responses in his study. The words, *rhetoric* and *hermeneutics*, are revolting to me now, and cannot be linked with healing. While I agree with his conclusion, his research results are dubious. How can one distinguish if actual healing occurs or the subjects have simply been *persuaded* to comply with the measurements of Flatland? As to hermeneutics, I repeat, no one can interpret the subjective world or symbols of another.

As previously stated, and worth repeating, even worse than simply being mistaken or misguided, both interpretation and persuasion can be designed to fit the hidden agenda of the therapist.

I can say, broadly, there are those persons in the field who *do no harm* and in fact, foster the healing process by disseminating knowledge, encouraging self-reflection, and the evolution of consciousness. Those persons are also aware of the effect of the client's current environment and of the influence of the socio-economic, cultural, and familial backgrounds on the client's development as well as traumatic events. It is certain those persons to not use their position of authority to oppress the client. While I believe those persons do exist, I have yet to meet one personally.

My experiences have instead, made me keenly aware of the validity of Toni Morrison's words: "Oppressive language does more than represent violence; it is violence; does more than represent the limits of knowledge; it limits knowledge.... Sexist language, racist language, theistic language—all are typical of the policing languages of mastery [control], and cannot, do not, permit new knowledge or encourage the mutual exchange of ideas."

Akin to Toni Morrison's words, *policing languages of mastery*, is *structural violence*, a term attributed to Johan Galtung, a Norwegian sociologist. It is violence against the human rights of another inherent in the basic structure of a system, as in Flatland. The field of psychology, like any other closed system, is fraught with such "policing languages of mastery."

Confronting all oppressive language worked its magic on me. With the encouragement of the *I Ching*, my dreams, and feedback from the first editor of my book, my words were getting stronger in my writing. In 2002 I was finally strong enough to confront the psychologist, in writing, for the harm he inflicted on me. I received no response. I filed a three-page complaint against him with the Florida Department of Health, Division of Medical Quality Assurance, detailing his actions. I received a form letter requesting information regarding specific laws that had been broken. I did not pursue the complaint; for I was confronted with what Toni Morrison terms *law-without-ethics*. There are no laws against his behavior.

Mental or emotional abuse was also finally acknowledged in the psychological community in the mid-90s. Apparently it does not apply to therapists—more evidence of a closed system that protects and defends the perpetrators, no matter how destructive.

I was then faced with another decision for my life. It has to do with revenge. Confucius' wise thoughts on the matter impressed me: "Before you depart on a journey of revenge, dig two graves." I had to let it go or remain a victim, a self-destructive kind of retaliation.

The contents of my book, *Of Golden Frogs and Such, a Story of Survival and Transformation,* begun in 2001, completed and published in 2006, took another year to assimilate. A year to gain perspective and to realize the writing was necessary to grieve through my nightmare experience and finally depart not only from Hannibal Lecter but the field of clinical psychology. It was another closure to the past.

Since then I have discovered there really is no need for revenge, for justice will prevail. Should that therapist and the thousands like him, wake up to the harm they have done, they will have to deal with their shame. Should they stay in denial, their spirits will never be free, and their souls will remain dark and menacing. Either way, they are pitiable, if not to be pitied.

I thought my writing days were over and turned to my music once again. My dreams, though, would not allow it, demanding that I begin again the process of self-reflection. I started a new journal in late 2007. As I was to discover, there had been some conclusions made in my book, *Of Golden Frogs and Such* that, through the test of time, proved to be limited interpretations in light of the unfolding future which has now become my past. I would like to provide a rational explanation for why I began this writing. All I can truthfully say is that it was time.

～

The I Ching was conceived thousands of years ago (some suggest early 3rd century BCE), initially based on the metaphorical laws of nature and Taoist philosophy. I believe its initial conception was truly an open system, but through the process of translation, it changed over time to reflect the bias of Chinese culture. Taoist Master Alfred Huang, in a retranslated English version, *The Complete I Ching*[63] published in 1998 delves into and reveres that culture, and ironically tells us the I Ching is essentially about evolution. He also acknowledges, and makes references to, the Wilhelm-Baynes edition I had been using for years, which remains one of the best English translations available.

For all those years, though, I was presented with the challenge to read around the yang and yin references to masculine and feminine, but no longer. I have discovered that the Tao symbol has nothing to do with gender. I have discovered, as well, that the edited and translated concepts of the I Ching are subject to the same errors as edited versions of *Flatland* or any other book—the problem of interpretation.

The structure of the human brain is essentially the same in men and women of the varied ethnic cultures all over the world, a basic structure that has not changed in over 40,000 years.[64] Yet everyone is unique. While some claim differences in male and female brains,[65] I contend there are as many differences between women and women, between men and men, as there are between women and men. The diversity is both nature and nurture, the multi-dimensional aspect of our lives.

The Tao symbol is a picture of two perfectly even hemispheres of the human brain with two eyes revealing the contra-lateral arrangement. The interactive quality of that symbol belies the numerous myths that split and divide us, especially a theory of opposites—simplistic Flatland thought. With its seemingly moving properties kept in perfect balance, it represents the wholeness so often sought and so painfully missed. Roger Sperry alludes to that wholeness with his term, *the emergent mind*, and our dual brains working in symmetry. Theodore J. Voneida[66] tells us of Sperry's concept.

> "The concept of emergence, according to Sperry, 'occurs whenever the interaction between 2 or more entities, be they sub particles, atoms or molecules, creates a new entity with new laws and properties formerly nonexistent in the universe.' He notes the parallel with quantum physics in which 'interactions among subatomic particles result in emergent properties which in no way resemble the particles from which they arose.'
>
> Thus, consciousness, in Sperry's view, while generated by and dependent on neural activity, is nonetheless separate from it. Consciousness emerges from the activity of cerebral networks as an independent entity. This newly emerged property which we call 'mind' or 'consciousness,' continually feeds back to the central nervous system,

> resulting in a highly dynamic process of emergence, feedback (downward causation), newly emergent states, further feedback, and so forth. Reducing consciousness to its separate components obliterates the emergent phenomenon of 'mind' with all its great power and uniqueness.
>
> Sperry elevated this concept of emergence from the individual to the global level, stating that 'the new paradigm affirms that the world we live in is driven not solely by mindless physical forces but, more crucially, by subjective human values. Human values become the underlying key to world change." (1972)

The *I Ching* also presents human values as an "underlying key" to its teachings, which I perceive to be truth, compassion, wisdom, justice, and dignity. Because **the** truth is rarely without question in evolutionary terms, I use the word *honesty*. *Kindness* is a personal preference over compassion, for it denotes observable action. What is wise is also questionable, as Abbot suggests, therefore I use his term, *the wisdom of nature*—evolution—that which contributes to our ascension to a higher level of consciousness and stands the test of time.

The entire structure of the *I Ching* is based on the creative (the yang force) that exists in each of us in our spatial brain and the necessity of our timely brains (yin force) being receptive to that holistic potential. While the book has its problems, it retains the essence of an open system when references to gender and social castes are removed; its philosophy acknowledges but makes no attempt to explain the Divine.

Such a book is a threat to closed system thinking. It was banned when the Communists took over China in 1949, for it was regarded as based on superstition and feudalism. It is frequently devalued and even ridiculed today in Flatland by those lacking in spatial hemispheric functioning.[67]

Admittedly, the *I Ching*, like any other book can be used in corrupt ways. It is not a divining tool for fortune telling, especially the trickery of so-called spiritualists. Since it does work through the unconscious and synchronicity, its metaphors offer counsel to the one seeking guidance. That guidance is for superior or favorable action in our daily lives and likely misfortunate consequences of inferior modes of action. Like precognition, it presents possible consequences based upon present conditions and existing information; those consequences can be averted by changing one's behavior—acts of free will.

Modesty is the most esteemed virtue in the *I Ching*. It is not to be construed as weakness or complacency. It is, instead, about the lack of pride and self-importance.

Modesty and the entirety of *I Ching* metaphors are about balance, not only the use of our dual hemispheres, but balanced lives. It is the excesses that produce personal disparities and the inequities in our human systems. Extremism, in any form, is not only considered to be inferior but an invitation for corruption, a concept applicable today and exemplified in unregulated capitalism or religious fundamentalism which evokes a terrorist mentality. There is an endless array of possibilities of how extremism affects our daily lives.

Master Huang speaks to another form of corruption when he states its teachings can be used for "mean purposes or selfish motivation." He is far too modest to say it has been corrupted and misused by Buddhism. That corruption is retranslation to fit Buddhism's archaic beliefs, condemning our emotional selves; also by omitting what Master Huang considers the most important part: the commentaries by Confucius, an honored sage.

While there is little doubt that Confucius was patriarchal, his wisdom was based on humanistic philosophy and the humane values that remain constant amidst ongoing change. He also exhibited modesty, that "most rare and excellent gift," to borrow Abbot's words, when he said at age 70: "If some years were added to my life, I would dedicate fifty years to study of the Book of I, and then I might come to be without great fault."

Editing out Confucius' translations or in some instances, the translations of Confucian scholars, is similar to editing out the teachings of another great sage and humanist, Jesus of Nazareth—the lost Gnostic teachings—which did not fit with the agenda of some early Christians. Even the word *Gnostic*, derived from the Greek word, *gnosis*, meaning knowledge, has been demonized.

I concede those teachings, like the biblical gospels, include the author's perception, hearsay many years after the fact, and the problem of translation. The similarities, though, between the hidden sayings of Jesus in Marvin Meyer's rendition of the *Gospel of Thomas*,[68] supposedly dictated in real time, with my own process of discovery, are nothing less than astounding. Meyer, regarded as a foremost scholar of Gnosticism, acknowledges that Thomas' recordings are very different from other Gnostic writings, for Jesus is portrayed as imparting wisdom in dialogue. There is no mention of physical miracles, fulfilling prophecy, apocalyptic warning, nor does he place himself above those around him, except as an example.

Regarding Gnosticism in general, Meyer says, "In Gnostic texts, unlike gospels of the cross, knowledge is more important than faith, and knowledge of oneself leads to salvation [recovery or healing]." Meyer's analysis of the Gnostic book of the Great Invisible Spirit and Secret Book of John results in his thought that the unfolding of the divine One is as much a story about psychology as it is about mythology and metaphysics: "the expressions of the divine One are mental capabilities: mind (nous), forethought (pronoia), thought (enoia), insight (epinoia), wisdom (sophia), even mindlessness (aponoia)."

Meyer's words reveal how man has, once again, created god (the divine One), in his own image.

For many years, even before my awakening, I shied away from anything that had to do with religion. I was vaguely aware of the Gnostic gospels but had no interest until I read *Beyond Belief, The Secret Gospel of Thomas*,[69] by Elaine Pagels, another highly respected analyst of Gnosticism. Her commentary emphasizes insight and creative thinking in *Thomas,* and she asserts the Gnostic writings, in many instances, reflect Eastern thought.

I do not disagree with Pagels. However, when I first read the actual translated Gospel of Thomas in Meyer's book, I found validation for many of my own thoughts and experiences, as well as its similarity to the fundamental principles of the *I Ching*.

The following excerpts are up front and personal:

"Let one who seeks not stop seeking until one finds; when one finds, one will be troubled; when one is troubled, one will marvel." (2:1)

"Know what is in front of your face, and what is hidden from you will be disclosed to you." (5:1)

"One who knows everything but lacks in oneself lacks everything." (67)

"When you make the two into one, you will become children of humankind." (106)

What the *two* refers to may be answered in Jesus' response to a question posed to him regarding entering the kingdom of heaven: "when you make eyes in place of an eye, a hand in place of a hand, a foot in place of a foot, an image in place of an image, then you will enter." (22: 6, 7)

Closely related is another maxim I find especially confirming:

" ... I say, if one is whole, one will be filled with light, but if one is divided, one will be filled with darkness." (61: 5)

Jesus, in *Thomas*, makes numerous references to self-examination. One of the more familiar ones: "You see the speck that is in your sibling's eye, but you do not see the beam that is in your own eye. When you take the beam out of your eye, then you will see clearly to take the speck out of your sibling's eye." (26: 1)

In contrast to the traditional *Honor thy father and mother,* Jesus said: "Whoever does not hate father and mother as I do cannot be a disciple of me, and whoever does not love father and mother as I do cannot be a disciple of me. For my mother gave me falsehood, but my true mother gave me life." (101: 1, 2, 3)

That pronouncement affirms the need to evolve beyond our parents and the parents we have become, the inheritance we carry from generation to generation. It speaks of the anger and sadness involved in closing the doors to the past as we become separate, discovering who we are. It speaks of the resolution and compassion that comes with understanding. It speaks of our true mother—the Spirit—that gives us life.

Another dictum especially significant to my story: "Jesus said, 'Love your sibling like your soul, protect that person like the pupil of your eye.'" (25: 1, 2)

Jesus, as portrayed by Thomas, was opposed to dogma that closes the door to new knowledge and gave few commandments. Another of them was "Do not lie." (6: 2)

Jesus was not prejudiced against women, for there is a general consensus that Mary Magdalene was his beloved apostle. In the Gospel of Mary, however, it is revealed how Peter, "the rock" on which the Catholic Church is purported to have been built, both distrusted and disdained Mary because she was a woman.

In *Beyond Belief*, Pagels presents a new historical picture of just how Christianity became what it is and its construction by the socio-economic, political, and cultural backgrounds (which included a long history of myth), as well as the personality of the pioneers and their personal environments.

Elaine Pagels and Karen King take it further in *Reading Judas, The Gospel of Judas and the Shaping of Christianity*,[70] to reveal the contention and disagreements in the early Christian movement. Their work disputes "what later historians depicted as an unbroken procession of a uniform faith," and exposes how those who gained power were "remarkably successful in silencing or distorting other voices, destroying their writings and suppressing any who disagreed with them as dangerous and obstinate 'heretics:'" the perpetuation of Flatland.

Essentially those founders of the formal church were successful in corrupting the true liberation that can be found in wholeness with a spirit that is free.

Much of Christianity has not served to enlighten but to entrap and control its victims through the enmeshment process, mainly through fear and the carrot of heavenly reward. One only has to explore its full history to discern the horrific persecutions invoked in the name of God to silence any voice in opposition. The contention and disagreement continues resulting in the splitting into countless factions. The oppression persists, essentially saying to women and to men, *Thou shall not be.*

I did not comprehend how entrapped I was until I began to extract myself from its dogma, so woven into my brain, a daily fear that lightning would surely strike me dead.

I can sympathize with Master Huang who, even in his attempts to be objective in his new interpretations of the *I Ching*, is still enmeshed in traditional Chinese culture. It is a culture that has been unable to evolve beyond its ancestor worship and patriarchy. Women, in general, have faired much better under communist rule. Yet that cultural bias—women are less than men and fit only for subservient roles—continue. Female infanticide marks its history, and selling female children as slaves, unfortunately, is not a thing of the past.

That Chinese (and Indian) tradition is also the source of one of my pet peeves—the insistence that the *posturing* of meditation is the only way to enlightenment and transcendence. My most profound insights have come from being in the world of nature, especially my garden. That was something specifically ridiculed in a book about meditation by an American Buddhist convert. (I cannot reference that book, for I threw it away.) The only thing we need to transcend is such closed system thinking.

Meditation, like yoga, Tai Chi, or any other relaxation technique is beneficial. None of them, though, is a substitute for dreams, for dreams serve as a gateway to our spatial brain knowledge; serve to bring synthesis to our selves, and to evolve consciousness. Perhaps meditation is preferred to subdue the world of feelings which become prevalent in dreams.

Of course dreams can be overridden and recall suppressed. There is also a plethora of information regarding controlling one's dreams.[71] It appears that self-knowledge is not of primary importance in either Flatland or Spaceland, nor has it been for thousands of years.

The storyteller in Genesis also condemned self-knowledge. It was a condemnation that evolved over time to *original sin* when "The

Church" projected its evil onto women and her offspring, then made it doctrine; sin from which one must be saved.

I knew that on Easter Sunday, April 15, 1990. I had written the following poem:

> Comes the dawn,
> rush to see
> a miracle
> from scattered thoughts,
> tattered dreams;
> no savior
> with neon signs
> flashing
> to enlighten;
> bitter thoughts
> from expectancy
> arises;
> sit quietly,
> put down the book,
> listen.

Shortly afterward I was sitting at my desk and heard a thunderous sound. I looked up and saw through the window in front of me, a large oak tree, a perfectly healthy tree, swaying from side to side. It was a still and cloudless day. Suddenly that tree snapped like a twig six feet from the ground at precisely 12 noon! It was an emphatic statement to cut man down to human size.

It was a statement that reinforced my task revealed in a dream on January 29, 1990, that I have alluded to several times. It was the dream that began my process; its contents appalling to me until I matured enough to realize its full meaning. The dream:

*I was walking alone beside an unpaved road, an open field on my left and on my right I could see the forest in the distance. In both hands I carried the feces of a baby. Then I felt a presence and put the feces in my mouth, hiding it as if it were shameful.*

To be as blunt as those shocking symbols: I had been forced to eat a lot of shit in my life, but some I did not swallow, including psychology's closed system premise (symbolized by Freud's psychosexual stages of development); but especially the doctrine of original sin. It was in my hands to expose the shame of that doctrine, for all children are born innocent. It was in my hands to condemn the marginalizing and castigating of women for eating the forbidden fruit of self-knowledge.

It took real time for me to pay the price of freedom, my path of self-discovery. It took real time to extract myself from closed system thinking and to gain the distance I needed to overcome the problem in closed systems, of not being able to see the forest for the trees. My capacity to see the whole picture also presented the task of bringing light to the darkness of that forest where poison mold continues to grow.

# VIII

# Complacency and the Heroic Antidote

The biggest waste of time is trying to figure out what is going on in the mind of another person. I may have intuition regarding someone, but I know I can also be mistaken. I am, though, no longer fooled by rhetoric, and if I watch what people do and listen long enough to their words, a theme or profile will emerge. It was another lesson I learned from clients and how I discovered my own patterns when I edited my biography.

Nor do I waste my time with those who **do** believe they know the subjective world of another or any "alternative religion" guru who claims to know THE truth and have all the answers to the existential questions beleaguering many. It is new-age dogma that can be just as oppressive as old-time religion, and in many instances is also based on the primal conditioning of reward and punishment. They are more high priests who have so many angles they diffuse into circles in Flatland.

When a guru begins to draw lists and numbered rules on the blackboard, I listen carefully. Eventually they will come to the "bottom line,"—the promise of financial reward—which is a reflection of what Toni Morrison refers to as "the bottomed out mind."[72] Or, like Ken Wilber, a self-appointed "pandit," who reduces human history and human functioning to his four dimensional so-called integral theory: individual objectivity and subjectivity; collective objectivity and subjectivity.

When I began to read Wilber's book, *A Brief History of Everything*,[73] I was impressed, perhaps because of my interest in the evolution of consciousness, and some of his psychological lingo was familiar. Then less than half way through, I realized he was engaged in a lot of *mystic, twinkling obsession*, dogmatic thinking, and exhibited no concept of modesty. Subsequent investigation into the "person" of Wilber confirmed my intuition that he indulges in the decadence of vanity and exhibits an inflated value of intellectual prowess. Despite Flatland academic savvy, his pathetic emotional immaturity is revealed in profane language and ruthless attacks on anyone who disagrees with him. He is just another tree in the forest that continues to oppress and darken the soul.[74]

Collectively, those new-age gurus join the rank and file of the complacent. The saddest part is the fallout: the countless followers who are lost in the forest, exchanging one religion, including the religion of psychology, or one cult for another, attempting to find themselves.

Complacency, the "vile and ignorant" land of the Point, was not an option for me, so I had no clue how one could become so immovable and self-satisfied. Nor did I realize how seductive that level is until I read the best seller by Rhonda Burns, *The Secret*,[75] which touts the power of positive thinking and not allowing a negative thought to enter one's mind. It is the epitome of the 3-monkey syndrome: see no evil, hear no evil, and speak of no evil. It also promotes an ancient and secret *law* of attraction: that we attract to ourselves everything that happens to us. I agree the collection of writings contained in that book is, indeed, an ancient story—of human greed—when the main attraction is money.

The so-called *law of attraction* is not scientific nor made by nature's law but instead its betrayal. It is the story of moral corruption. The extreme of that corruption is when a man suggests, as Joe Vitali does in that little book, that the holocaust or any other mass bad fortune is "attracted" by some inner condition of the victims. No different from Pat Robinson, a fundamentalist Christian televangelist who attributes disastrous events, e.g., 9/11/2001 destruction of the twin towers, as God's wrath.

Such concepts are as horrifying as the story of *Neighbors: The Destruction of the Jewish Community in Jedwabne, Poland*; as horrifying as the holocaust, Darfur, or any other human-made destruction. The holders of such concepts essentially become participants in those crimes against humanity, as do the mindless followers who adopt the rhetoric and give tacit approval.

Just as gurus cannot exist without gullible followers, Flatland could not continue to exist without the conformists, both the product and producers of the fog. They are the individuals hiding behind closed doors and living in a perpetual state of denial of their own inner world and the world around them. They are so adept at creating their own story of perfection, adept at creating their own reality, and believing it so strongly, an authentic person is not there. They can be masters of entrapment, masters at pulling another into their story and seem to be totally oblivious to the effect they have on other people; it is the people around them who suffer. They are truly deaf, dumb, and blind, bereft of sensate existence.

They are the "red flags" spotted throughout my history when I was pulled into their stories, including my so-called therapist. He was such a *nice* man I put my intuition in override. His pacifist posturing—Quaker identity—became ludicrous, though, when years after my experience, I, along with others, witnessed his rage. Anger is often a part of the grieving process, as is denial or sadness, and some become stuck in those various stages. Like many other therapists I encountered, he professed that grief is a waste of time. He went even further with one of his dogmatic statements: "Anger is the basis of neurosis," and it probably is. But it is the denial of anger (often becoming the slow-burning fire of resentment), that is at the root of the dissonance with which he has to live.

Even though his father was physically abusive, that his rage was directed at women was no surprise considering another one of his statements to me, "You may as well direct your anger at your mother, for that is where it always winds up." It is the subliminal stance that begins so early it appears genetic. His familial history he shared with me suggested it was also a projection, for his "difficult" mother, at that time, was still sleeping with his adult brother. Such dishonesty with himself is suggestive of a divided self and essentially, absence of soul.

The word *soul*—the "I" behind insight and the unifying center of one's being—is a word bandied about loosely. Like a woman discussing home décor on the Oprah Winfrey show said, "My home is the reflection of my soul." It is the kind of mindless statement that speaks of the shallowness and blindness of Flatland. I wanted to ask her, "What about the homeless or those living in shanties or tents? What do their souls look like?"

I also thought of the popular myths that support her mindless statement, especially the totally erroneous concept that art reflects the soul or spirituality, for even "pickpockets and cutthroats" are apprised of Spaceland and creativity. I wondered if she had been reading those Freudian dream book analyses and "archetypal symbols" purported by Jung, particularly the conception that house dreams represent the self. And it frequently does. As said before, it represents the self that has been constructed by humans with all its numerous rooms (compartmentalization),[76] and a house we must leave if we ever expect to find our wholeness and our true selves.

Another myth maintained in Flatland is that evil is a necessary and inevitable opposite of good. Alice Miller, in *Banished Knowledge*,[77] addresses that myth and conceptualizes it in a way similar to the ancient wisdom of the *I Ching*: "... evil is real. It is not innate but acquired, and it is never the reverse of good but rather its destroyer. It is not true that evil, destructiveness, and perversion inevitably form part of human existence, no matter how often this is maintained. But it is true that evil is always engaged in producing more evil and, with it, an ocean of suffering for millions that is similarly avoidable."

Children **are** born innocent and unique; our multi-dimensional lives present enough challenge without evil, for perfection does not exist. It is the luck of the draw, the family and environment we are born into, as are our genes that dictate the formation of our bodies. There are chemical or structural weaknesses and anomalies that challenge many of us. Such is the case of Bipolar Disorder, a medical condition with an underlying chemical and neurological structural anomaly which does not deserve the label, *mental illness*. That label presents a social stigma compounding the challenge of the illness.

In my case, I have scoliosis, an S-shaped curvature of my spine that threatens to progress, just as my mother's did, and numerous herniated disks in my back and neck. At times I have experienced excruciating pain. My condition is another incident of our accidental universe where bad things just happen, like the child who has cancer, or those who are victims of the poisons in our environment, or ...; the list is endless.

New-agers, in many instances, have become extremists with their "mind-body connection" by asserting illness is punishment for some unknown crime, just as ludicrous as the term *god's will*. When physicians, counselors, chiropractors, massage therapists ... view all illness as having an emotional cause, it is just a different spin, but same concept, as doctors assessing many authentic illnesses in women with, "It is in your mind." No doubt related to a categorical statement by one of my psychoanalytic professors in graduate school: "Lower back pain, in women, is a symptom of neurosis."

With my health condition (and no health insurance), I have had encounters with many of these fanatics. For years I remained silent. My experience with my so-called therapist taught me that no one knows what I feel or think and no one can tell me what I am supposed to feel; neither am I an object for anyone's assessment.

One such fanatic encounter was with a female chiropractor who had a one-size-fits-all approach to treatment. It was after a year of ineffective intervention when she said, "I wonder what you did in a past life to deserve your pain." When I objected to her statement and explained my scoliosis as genetically acquired, her retort was, "But you chose your parents." I confronted her B S in the moment and later wrote her a letter detailing the unquestionable destructiveness of her words, especially that how she "sees" someone affects the way she treats them. I was astounded at her response: "I would never say such a thing to anyone!" then began to verbally attack me.

It is an exemplary example of another divided self, when her right hand did not know what her left hand was doing. The sad part is that she was, in all probability, not consciously lying and actually believed her ego centered story—another darkened soul.[78] When I confronted my former over-the-edge, new-age physician (who said, "Bad things do not happen." yet associates pain and suffering with the lack of forgiveness), the response was just as pathetic.

One of the most bizarre encounters, though, was a massage therapist who adheres to a "feeling" map of body parts with each representing a particular emotion. His perception was so ludicrous and mindless, I did not waste my time to confront him. Our emotions, despite romanticized words such as *heartfelt*, as well as our thoughts, emanate from structures in our brains.

By breaking my pattern of dependency, I have become more physically *aware* than most people ever think of becoming by researching and gaining knowledge about my body. While chiropractors facilitate structural balance of the spine, my condition happens to be a muscular deformity resulting in *functional* scoliosis that did not manifest until midlife. Through experimenting with various methods of treatment, I have designed a unique program that works for me: walking, yoga stretches, trigger point self treatment, relaxation, mild pain medication, and therapeutic massage by a kind, non-judgmental woman. Physicians are amazed at my physical fitness, in addition to the fact that I am not in a wheelchair.

There is no question that we can neglect our bodies and, through extreme behaviors, cause all sorts of physical problems. Nor is there a question that stress affects our health by weakening our immune systems, making us more vulnerable to illness. Our health can also cause stress until we accept our limitations, take responsibility, and make peace with whatever our luck-of-the-draw has given us.

With some things, though, we should not make peace and one of them is complacency in any form. Nor should we make peace with the guilt-provoking, destructive, and demeaning rhetorical dogma from any extremist. It is oppressive language that does violence to others and language revealing the speaker, in spite of any pleasing façade.

༺༻

In contrast to conformists of Flatland and the *new-age* religious alternatives, there are true heroes of moderate and modest minds who do not gain popular approval and are rarely the subject of news media interest. They are heroes who have found their purpose in search of truth, justice, and work to evolve major religions to a more humanist form. They understand the necessity of looking backward in time to expose the roots of violence and oppression in patriarchal religions and cultures all over the world.

The beginning of that movement was in the 50's and 60's in South America by Catholic priests whose focus was on the suffering of the poor and disenfranchised. *Liberation Theology* is the term used for that movement, but much of its work has been thwarted by the Vatican that condemned it as being Marxist. Its ideals, though, have now been used by African Americans and women in a struggle for liberation from the fog of Flatland.

I can relate to Liberation Theology, personally. During my tenure working for the Episcopal Church, it became evident that the rhetoric and trappings of rituals were simply that; behind the scenes were the political splits and divisions within the church, the inequalities, the misuse of funds, the sexual indiscretions—widespread hypocrisy. Even though I left the church and organized religion when I was 35 years old, I was unaware of the residual effect of traditional Christian brainwashing, especially the use of fear.

Yet I had found two *liberating* statements within the traditional Christian doctrine that influenced my life. They were, paradoxically, the foundation for my eventual emancipation from the confines of organized religion. One was recorded in the Gospel of Matthew:[79] "Truly I say to you, inasmuch as you have done it to the least of these my brothers, you have done it to me." The other was similar to a saying in the Gospel of Thomas: "Seek and you will find; knock and the door will be opened to you."[80]

Daniel C. Maguire and Sa' Diyya Shaikh, in their book, *Violence Against Women in Contemporary World Religion*,[81] provide a collection of writings by some of the modern day heroes addressing each of the dominant religions within the context of their own cultures. What I find so heroic is they seek the cures within the religious teachings. Once again, it is a matter of interpreting or re-interpreting the language.

I must admit I am limited, for I am not so tolerant, and I find it is really stretching it to reinterpret patriarchy as anything other than patriarchy.

In the Maguire and Shaikh book, several opinions about the origins of patriarchy are offered. One of the theories suggests the myths and legends preceding biblical myths originated when man began to believe he was master over the earth, subjugating women and each other. In which case religious doctrines were formed to legitimize and attribute to god the bias of the culture. Perhaps like civil war, written into the book of Genesis, when that storyteller's masculine god—made in man's image—preferred the offerings of animals to the fruits of the land, pitting brother against brother. It seems to be a story legitimizing the civil war that continues to this day in the Middle East.

Another theory of the origins of patriarchy is by anthropologist Peter Farb,[82] and alluded to by Maguire, that suggests women needed protection from predatory men and relied on their mates for that protection, creating a social dependence. But social dependence does not account for the violence against women and men, essentially condoned in many religions. I find Isaac Asimov's thought is applicable: "Violence is the last refuge of the incompetent."

Violence is, inevitably, hidden within closed systems. We now live in an information age, and at least we are no longer victims of silence. The contributors to Maguire and Shaikh's book are but a few who expose not only oppression but the criminal violence, both sexual and non-sexual, perpetuated on women, boys, and girls, every second of the day in every country of the world. That includes the U.S. Although they may not be visible in mass, they number in the millions, those living tortured lives or becoming the walking wounded. Abu Ghraib is not an anomaly. I repeat, I, personally, am limited, for I no longer have tolerance for closed systems. Yet I know people can only adjust by degrees.

While subjugation of women, sexual abuse, and harassment occurs in every religion and in every denomination of Christianity, the Catholic Church, by retaining its medieval doctrine of celibacy for its priests, is an open invitation for violation of vows, sexual deviants, and misogynists.

Pope Benedict XVI's addressing the exposed sexual violation by priests in his 2008 visit to the U. S. is laughable. Finally, that scandal is being revealed as merely the tip of the iceberg. He, like his predecessor, through his demand for silence, has served to perpetuate the covering up of the rape crimes, not only on children, but women, including nuns, bribing and intimidating victims, and protecting the abusers. His cursory statements, without owning participation, do not give the details of the devastation to the lives of the wounded. That high priest, who has been sold a bill of goods about his holiness and infallibility, fits the profile of another emperor with no clothes!

I am the first to admit my passion is personal, not only because of his denial and his sexism that rejects my human right to make choices, but my companion was one of those damaged in his youth. Like many other victims of rape, he blamed himself.

My companion was a man who exhibited well-developed linear and spatial brain abilities with a love for music matching my own. Yet, as previously stated, we encountered a repetitive impasse in our relationship when he would withdraw into his world of silence. There was hope for us when he engaged in therapy with Hannibal Lecter: 10 months of supposedly working on communication skills, addressing the symptom, not the cause. Concluding his *therapy* was that catastrophic marital session when I was *invited* to return alone. Even after my fracturing experience, we tried to pick up the pieces.

My real hope for us was when he began to share his inner world with me, notably his dreams of the symbolic demons—the sources of his fear—that had been chasing him for thirty years. The demons that were human made by the violation inflicted on him. He also began to report unusual moments of insight, well on his way, I thought, to his own emergence. His commitment to follow his feelings, his love for me, was what I had longed for.

After I joined him in Poland, it was not long before I witnessed his relapse to a person I did not know. The only clue I had were his words, "I am not strong enough to do what you are doing." All the gains he had made began to dissipate as he gradually became re-enmeshed in Poland's Flatland, especially fundamental Catholicism with its sexist and anti-Semitic prejudice. I watched as he retreated into his world of silence and obsessed with work, drinking, money, men.... He seemed to lose any personal integrity and any touch of kindness even for me.

It was painful to realize that my companion, whose emotional development was thwarted at an early age, was not only enmeshed in closed system thinking, but his soul was relegated to darkness. Our time together was over.

It was man who created hell on earth and evil, and it was man who invented a word to excuse it. There is no more misused, enabling, and disabling word in the English language than *forgiveness*. It is often used to bypass anger and sadness, the pain of grief. It is a word discouraging the healing that comes through understanding and the closure that comes with resolution. It is a word that feeds, instead, the 3-monkey syndrome—complacency.

*Forgiveness* also often suggests there are no consequences to what we do, but there are; more frequently than not, those consequences are visited on others and carried from generation to generation. Those collective consequences have accumulated over time to the reality of the here and now.

The governing body of our patriarchal Flatland is and has been in excess of 99 percent religious, retaining the motto, "In God we trust." Yet it has sanctioned a consumerist and materialistic country where the rich have become richer and the poor have become poorer. Many people live in abject poverty. It is a country living on borrowed money and borrowed time; and in its financial crises is reaping what has been sowed. Instead of being a shining example of a great country, we have lost credibility with the world.

It is a country where honesty seems to have become a commodity and kindness the exception, not the rule. Justice wears a blindfold and is blind to human rights. That materialism, in its inhumanity, now infiltrates the food we eat, and it is human waste that pollutes the water we drink, the air we breathe; not to speak of the deadly nuclear waste hidden beneath the surface of our earth or the oil-spill disasters. Safety is an illusion. There are real threats to our society and the causes of most of them are rooted within our society.

There are, however, those countless heroes who are not deceived by rhetoric and have found their own sense of purpose in exposing the corruption that has accumulated in the multiple dimensions of our lives, serving to evolve awareness. They are actively engaged in fighting to save our earth and its endangered species as well as social justice. They are actively engaged in removing the state of denial on which evil feeds.

One of those true heroes is anthropologist and physician, Dr. Paul Farmer, who gives Liberation Theology clear meaning and is an expert at elucidating the structural violence that permeates closed systems. He is a man who, by his life and his work begun in Haiti (extending now to many countries), reveals how one person can make a difference in the world. My daughter, engaged in ethics as a professor and in her daily life, introduced me to Dr. Farmer's work by recommending Tracy Kidder's Book, *Mountains Beyond Mountains*.[83] I wept my way through the last chapters when the stories told about the poorest of the poor in our Western Hemisphere moved from abstraction to reality.

I have a sense of brotherhood with "the person" of Paul Farmer—his words, his actions, and his history. He came from the same "neck of the woods" as I did, and I identify with his passion when he referred to himself as "Mother" to the people he doctors. There are numerous incidences described by Kidder suggesting Farmer is aware of the complexities of survival and how it plays a major role in human behavior.

Then when I read Paul Farmer's own words in his book, *Pathologies of Power; Health, Human Rights, and the New War on the Poor*,[84] I had the same response I had in reading the Gospel of Thomas. I found myself jotting down his ideas that validated my own insights until I realized I would wind up rewriting all of his concepts. While Farmer's focus is the human right to have medical care and humane living standards, he elucidates the structural violence permeating our human systems that is a betrayal to those rights.

His concepts are applicable to all oppression and reflect, only different language, the closed systems of Flatland with built-in pathology. It is definitely applicable in the field of mental health. Structural violence is inherent in the labeling and measurement of human worth and became obvious to me in the community mental health centers where I interned and acquired a reputation as *client advocate*. It was even more obvious in a Drug Treatment Center where I worked years after my awakening (See Appendix III). Those who need care the most, generally receive the least in quality care.

No summarization or interpretation of Dr. Farmer's writing would suffice to reading a book Robert Lawrence justly says, "captures the central dilemma of our times—the increasing disparities of health and well-being within and among societies. . . . *Pathologies of Power* makes a powerful case that our very humanity is threatened by our collective failure to end these abuses."[85] It is a book that does explore socio-economic, cultural and political causes and elucidates, particularly in Haiti, the role the United States, along with other countries, has played in the destruction of countless lives.

Farmer makes a clear distinction between social justice for all and the quests for power sharing by formerly disenfranchised groups. He says "The identity politics of our times has a troubling subtext: *I've been wronged in the past, and I want what's coming to me.*"

In contrast, Farmer refers to Brazilian sociologist, Paulo Friere, who believes in the necessity of his coined term *consientization*—the deepening of the attitude of awareness characteristic of all emergence—to understand how social structures cause injustice. Perhaps it is another way of saying that the "attitude of awareness" requires the leaving of Flatland and moving toward a major awakening—the evolution of consciousness—to see and to change those structures.

While I was reading Farmer's book, another extraordinary man's name kept blipping into my mind: Nobel laureate Albert Schweitzer, medical doctor, theologian, philosopher, and musician. Reading James Brabezon's biography of Dr. Schweitzer[86] was eye opening in the light of structural violence and emergence.

Brabezon tells us that Dr. Schweitzer was disturbed by what he considered to be the collapse of civilization in early 20th century Europe, the lack of ethics or humanitarian ideals, and had been ruminating on a worldview answer to his question, "What is civilization?" He had found no satisfactory answer in religion or philosophy and had come to believe he was in practically unexplored land. Dr. Schweitzer's own words were used to describe a moment of insight while on a boat trip in Africa when the "iron door yielded." "There flashed upon my mind, unforeseen and unsought, the phrase, 'Reverence for Life.'"

Brabezon considers those words as a culmination of Dr. Schweitzer's "having felt the force of continuing life in the vastness of nature." The author also tells us that when that phrase came to him, Dr. Schweitzer did not think of Christianity or philosophy, his mind went to eastern religion, for the antecedents of his great idea were, indeed, worldwide.

Dr. Schweitzer elaborated his insight: "Ethics is nothing other than Reverence for Life," and "Reverence for Life affords me my fundamental principle of morality, namely, that good consists in maintaining, assisting and enhancing life, and to destroy, to harm or to hinder life is evil."

James Brabezon presents his own understanding of Reverence for life:

> "Reverence for Life says that the only thing we are really sure of is that we live and want to go on living. This is something that we share with everything else that lives, from elephants to blades of grass and, of course, every human being. So we are brothers and sisters to all living things, and owe to all of them the same care and respect, that we wish for ourselves."

I was surprised to learn in Brabezon's book that Albert Schweitzer was criticized for his patriarchy and did not see the African as equal but a *junior* brother. It is further suggested that Schweitzer also treated them like irregulars, assigning meaningless tasks to them.

In Dr. Schweitzer's defense, he seemed to adjust by degrees from the Patriarchy so engrained into his life, for after his insight, he stated that it was time to consider the Africans no longer as junior. He was well aware, nearly a century ago, that they were victims of structural violence inherent in slavery and European Colonialism. He suggested, even then, that it would take time and perhaps generations for their cultures to overcome the inequities that, unfortunately, continue to this day.

I consider Dr. Schweitzer's *Reverence for Life* insight at the least a "peak experience" and just as likely a glimpse into the soul—the unifying force—where true universality exists.

## IX

## Make Friends with Time
## One can only adjust by degrees

Just as I found the seeds of liberation in my Christian teaching, I also discovered, within the system of psychology, the seeds for my liberation from that system and the tribe mentality into which I was enmeshed. Psychiatrist Murray Bowen, a family systems theorist (as well as Alfred Adler), considered enmeshment as an impairment to his perceived life goal of individuation, for enmeshment blocks the process of becoming a separate individual, thwarting one's capacity to think, feel, speak, and act on one's own. Individuation, freeing oneself from that entrapment, is central to both Bowen's and Carl Jung's theories. It is a concept that intrigued me as had Abraham Maslow's apex of hierarchical needs, self-actualization, the process of bringing actuality to the potential of the self.

In all probability none of these theorists were aware of the numerous dimensions involved in achieving their theoretical ideals, especially the role our dual brains play in those processes. Nor were they aware of the role neurobiology plays in emotions, learning, and memory or the neurological changes that do occur with experience. I know from my own process of change, that old modes of thinking, feeling, and doing, were more than simply habits, for they only diminished as I replaced them. All of which is to say, it is a long way from insight to change that is permanent.

I have relapsed into old behaviors many times, especially self-doubt and dependency, which are essentially the same. Transitioning from enmeshment *theory* to full awareness of how easily one can be pulled into another's story was, indeed, a process of adjusting by degrees. I was also still learning to identify and acknowledge my true thoughts and feelings. Like A-Square, for many years I experienced the feelings I had been taught, squelching those that were truly mine, just as I had suppressed spatial brain activity out of fear. It is likely they are the same suppression for our unconscious—our spatial hemisphere—is the source of authentic feelings, emotions that are not based on reason.

It is well known in the field of psychology that one's subjective world, often containing the baggage we carry, can distort our perception of the world around us in the process called *projection*, a word I have often used in this writing. It is a process that is one of the sources of illusory existence when one ascribes to the outside world, such as an image or other people, one's own subjective world.

Projection is also the basis for the notorious *Rorschach* (inkblot) test, and many other psychological tests that, as said before and worth repeating, have been standardized by comparing an individual to the performance of a larger group composed, predominately, of Flatlanders. Some projective tests make use of art in the form of drawings, which often produce spatial brain knowledge in the form of symbols which are then *interpreted* by the psychologist. Testing is a lucrative business, and those tests are not used therapeutically, but to label—dehumanize—the client.

Even with the knowledge that drawing taps the unconscious, the majority of the high priests of the psychological community has failed to embrace art therapy; that would entail departing from the dogma of Flatland and the chance that, with the "use of color," all would appear equal. Patients might even discover the means for healing themselves.

It was something I had to learn on my own as I worked with clients. Betty Edwards validates that concept with her words: "Drawing can reveal much about you to yourself, some facets of you that are obscured by your verbal self."[87] She also suggests learning to draw and drawing requires giving up left-brain control and allowing the spatial hemisphere to take charge. It is a process of learning "to see" what is actually in front of you instead of images distorted by Flatland reasoning.

But it goes even further than just learning to see. In my own case, it was shocking to realize long before 1990, my story—past, present,

*AND* future—had already been told in my paintings and drawings, those things I knew but did not know I knew.

In 1975 my first college course was applied art. The assignment was to choose photographs and reproduce them in oil media. One of my choices was from late 19th century England, a waif alone on the street and overshadowed by the images of a cathedral in the background. I inserted, however, a spirit form that was with her like a ghost emanating from the top of her head.

When I painted that picture, something about the process frightened me, for it threatened that *intricate illusion of sanity*. I falsely believed there was a line one could cross from which there was no return. Of course it was the Other World of Spaceland which is, indeed, a threat to patriarchal society and its egoist mentality. I did not paint again for a decade.

Then one April day in 1985 I felt compelled to draw and produced three pictures. The first one revealed the binding around my head—my emotional dependence—in my relationships. The second drawing was a self-portrait, a woman enclosed in a cocoon-like structure, the chrysalis from which the real me was to emerge. The story told by the third drawing, another self-portrait done in pastels with no semblance of cover-up, portrayed foreknowledge of the emergent mind.

When my poetry began in April 1990, I did not understand the source and experienced moments of fear that those poems might be stolen from me. As mentioned previously, they came as though channeled through, not from me. But they were from me, my soul, breaking through the illusions of Flatland or Spaceland consciousness. They came in my dreams, during conversations, sometimes, for lack of better words, out of the blue as described in one of the earliest ones:

> Word aphasia,
> even brained,
> an oddity;
>
> A million thoughts
> locked inside,
> no words could tell;

They tumble out,
and the taste
is bittersweet.

Those words did continue to tumble. Their arrival was unpredictable as was the content. I would count the syllables to find the form. They came every day, sometimes twice, in April 1990 until my so-called therapist grabbed my journal of poetry and threw it on the table. It would be 10 years before I realized those poems were to be my guide in my transformation.

One of the most important poems came in a dream; I awoke and simply wrote it down:

Heavenly bodies of childhood lose their brightness
with spoiled dreams and books
and close-up photographs;
a caricature life becomes, like heavenly bodies.

You can take heart if you want to marvel,
one phenomenon grows with each encounter,
the human spirit,
greatest mystery of all.

Appreciate the complexities of survival,
for no spoiled dreams or books
or close-up photographs
can take away the wonder of that ascending power.

My understanding and interpretation has progressed over time. Children are born innocent, totally open and vulnerable, with the instinct to survive. Their dual brains are initially, undifferentiated. During childhood, the brightness of curiosity, creativity, imagination, and intuition are simply there. The introduction of fear to those innocents comes from their environment, often feelings too overwhelming to cope with. Feelings relegated to dreams and the baggage carried or acted out symbolically.

That spatial brain activity is trained out of them with the books of the three R's in our educational systems, the influence of cognitive developmental theorists like Jean Piaget—left brain bias. Close-up photographs are the criticisms and measurements of who or what they are—diminishing labels and judgments—the reductionism that is the pathology-based field of psychology and the fear-based field of religious mythology.

Without that reflective side, and the insight it can offer, life does become but a caricature in Flatland, living the illusion of sanity. It is the mechanistic view which is in fact, insane, for the same mistakes are repeated over and over again. The world of robots, referred to by Asimov, as a possibility for the future, is already here.

Encountering the human spirit does evoke a sense of marvel. The spirit, the life-giving energy of the space-time-matter-energy is simply mystery. We do not know what it is, and it is difficult to grasp its extensive and mysterious properties. As we are beginning to comprehend, we do not know the limits of its power.

The complexities of survival are exactly that, and one can only come to understand those complexities by self-discovery and maintaining a longing not only to survive but to evolve.

What I failed to comprehend for a long time is that the soul's voice often has universal portent.

෴

We will not move out of the myths that close our human systems until we wake up to the fact that we are, inextricably, a part of nature. Both accidental and orderly, nature, in its wisdom, provides the structure of the seasons, the structure of the day, time that existed long before humankind, yet man made calendars to accommodate nature's patterns, a process of discovery.

We have evolved enough to discover there is order in the seemingly chaotic profusion as life took form and even in the formation of the surface of our planet: "phenomenally elegant recurring mathematical and geometric patterns known as fractals." In nature, for instance, a tree branch or a frond from a fern is a miniature replica of the whole: although not identical, similar in form. It is likely that human life is also part of the fractal order, and I propose a child's development is similar to the pattern of human evolution.

Conceived in a sea of nurturing fluid, developed to a crawling then walking stage, and like an infant, the human brain was initially primarily intuitive. As the dual hemispheres evolved, they were, like a child's, undifferentiated by language, and symbol production was dominant. It was art that evolved into the written word, and there is evidence that music preceded both. The "Word"[88] gave man the illusion of consciousness and a sense of power and control. He progressed to condemn what he could not control, losing, in the process, his wholeness and his connection with nature and the unifying source of life—his soul.

One change in the pattern, though, can change the whole order of our human systems, the reinstitution of "color," a new Renaissance (the period of time Abbot referred to, when all appeared equal). An awakening of the holistic potential that exists in each of us would change our black and white world of Flatland. And with the amazing diversity constituting humankind, in the mutual sharing of ideas, one can conceive of evolution of consciousness to increasingly higher dimensions as encouraged by Abbot—in time.

The seed of idea takes time to germinate, and as a people we must tend another long-lost attribute as dictated in another poem:

Do not contemplate on confusion;
go to the garden to see the art of creation.
I cannot a big tree bend nor make flowers bloom,
only clean my own debris, plant a seed,
and tend the art of patience.

So many times we are taught that in the whole scheme of things we are insignificant. Because our gifts seem so small, we do not give credence to them and, through carelessness, throw them away. Just as we think one little fault does not matter, like a weed it grows and multiplies, overtaking the beauty of the human potential. It is through changing reductive thoughts about ourselves we can discover our own worth. Gardening is not an easy task; it takes time, effort, and patience. In order to clean one's own debris, the first step is self-examination.

While there are many ways to examine one's life, I cannot expound enough on the value of writing and the transformation it has brought to me. The rough drafts of all of my works occurred around 4 a.m. when I recorded unedited thoughts and ideas on paper, essentially a spatial brain activity—the whole picture—in words thick with meaning. Editing, for me, is a linear brain activity of expanding those images or symbols in words. The tedious process of providing details often brings increasing clarity. There are numerous times, though, when a dream or waking thought will direct me to make corrections.

The process of writing both my first unpublished book of personal history and my second book, *Of Golden Frogs and Such*, involved becoming painfully honest with myself and the inevitable grief. That pain, however, was compensated with profound joy as I began to discover my authentic self.

This writing has been, predominantly, a process of discovery, exciting and increasingly freeing as I integrate those discoveries. Finding validation in the thoughts and experiences of others has been priceless in moving what was initially intuitive to reality. That validation is empowering, and I have gradually learned to trust my own voice.

Peter Elbow, in *Writing With Power*,[89] alludes to some of those same processes when he talks about why people do not use their *real* voices, and one reason is "that it means having feelings and memories they would rather not have." He contends writing in real voice language often brings: "tears or shaking though laughter too. Using real voice may even mean finding you *believe* things you don't wish to believe." Elbow also tells us that some writers acquire *real* voices through fantasy, as in the case of fiction, and through lies—the convincing rhetorical propaganda to which we are all subjected. As he says, though, "Many good literary artists sound least convincing when they speak for themselves." Then, of course, there are the "memoirs" of some rich or famous that requires no voice at all except the concoction of ghostwriters and editors behind the scenes.

While not all *truthful* writing brings transformation, there are some who have experienced such change. The difference seems to be those who also worked on their own debris and/or opened their minds to something larger than themselves. A recent testimony can be found in *The Freedom Writers Diary: How a Teacher and 150 Teens Used Writing to Change Themselves and the World Around Them*, by Freedom Writers and Zlata Filipovic. It is the basis for the 2007 movie *Freedom Writers*. It is also another example of what one person, in this case, a teacher can do. It reveals how opening minds can bring about transformation, challenging our educational system's Flatland orientation.[90]

Patience was often trying—the reality of the word *gradual*—but the hardest lesson I had to learn and one that was another source of much grief and loss: I cannot change another. I once heard there are three kinds of people (of course there are more): those who see a problem and set about to find ways to fix it, those who see a problem and do not know what to do. They are the ones in whom one can take hope. Then there are those who say there is no problem. That lesson was dictated in another of my poems:

Do not bother with those who do not seek,
hiding behind the scene of perfection.
Their pointing fingers show
the wrong direction.

Although we cannot change them, we can bring their misdeeds to the light of day and remove the food of darkness on which evil feeds.

Many of the concepts addressed in this writing, especially the closed systems of religion and psychology, patriarchy, and structural violence, are concepts that fit with "there is no problem" by the complacent majority. Early in my process, some of my challenges were met by adages frequently repeated to myself, a simplistic but effective way of encouraging my own changes. One was, "Dare to be imperfect and keep planting seeds." Another came from a favorite childhood story, "Believe in the little engine that could." Then the Disney portrayal of the inchworm: "Inch by inch life is a cinch; yard by yard life is mighty hard for an inchworm."

Another favorite bit of wisdom is found in a furry little friend: "Sing, sing a song. Make it simple to last your whole life long. Don't worry that it's not good enough for anyone else to hear, just sing, sing a song."

One of the strangest poems written during April 1990—just out of the blue—still makes me smile:

> I'd like to write a pretty poem
> of flowers and light perfume.
> Until my entire tale is told
> I'll have to live with gloom.
>
> Sometimes I laugh inside myself
> at the antics of my doom.
> Perhaps this tale turns out to be
> of flowers and light perfume.

It seems to fit with one of my paintings done in 1998 titled *Hope*, an image feeding my visual need for reassurance. It is a face emerging from the darkness of grief with a vision of a new kind of tree: one not so tall, shaped by weathering every storm, and still stands because its roots are firmly planted. It is a tree that blossoms and bears fruit.

For a long time the painting was somewhat disturbing because the face of the woman is young. I knew I was not in a state of denial about my age but did not discern its meaning until my mind's eye kept seeing an image of me as a youth. I had been sent to the field, a mile-trek from the house, to work alone. It was an image of loneliness, isolation, and the despair inherent in being "different" with so many longings inside—a Spacelander in Flatland. No human activity was in view, just the sound of trucks and cars on nearby highway 50.

It was my task to hoe the weeds from the acres of beds of seedlings. Up and down those rows, one after the other, weeping as I trudged along for the duration of the day. It was a time when I was truly open and innocent. That painting was symbolic of a concept surpassing anything I had been taught—one can return to a state of innocence. It begins by freeing one's spirit from the "clutches of obsession", redeeming one's wholeness, and discovering one's soul.

It once was thought that the amazing human brain is fully developed at the age of 9, an age that has been extended to 25. Now I know our brains can continue to develop throughout our lives. It is the truth: people can and do change by maturing and evolving. No matter our age it is not too late or too early to start.

The first thing, though, is to get our egos, the Flatland beliefs about ourselves, out of the way (Joseph Campbell calls them the dragons we must fight), and be receptive to the holistic potential. In the same train of thought, sometimes I do not copy forms in my art but simply go to the canvas and allow my spatial hemisphere full expression as forms take shape from the color. What the creative in me produces continues to amaze me, for those paintings, like my music and poetry, are a product and producer of the "greatest mystery of all."

## X

## Embrace the Mystery

I had to go through the process of mending the human-made splits and divisions in my mind, releasing myself from the chains around my head; the chains forged by belief and fastened by fear. I had to go through the process of becoming whole to realize I had always been whole and had been offered many gifts (the human potential); gifts that were up to me to use or not.

Many times I have been frustrated with people who say how lucky I am to be creative. As suggested by the process of learning to write, none of my creative processes have been quick or easy for me, especially my music. I did not consider myself a musician in the traditional sense, but my teacher encouraged me to bring my musical poetry to fruition. He said, "Composing music is about 15% talent and 85% fortitude." My music, though, was more than simply compositions; it spoke of more things I knew but did not know I knew.

The most formidable task of my process has been to trust my insights regarding the soul. While my poetry, art, and writings provided feedback through my senses, it has been my music, the universal language, which has served to unify all the disparate parts of me and to validate my center, my connection with all that exists. It has served to validate *essence* emotions. I play it often to remind me of who I am and my sense of purpose. It is a purpose born in the sacred love for my brother and my daughter. It is a purpose that became emphatic with the dream initiating my process—the innocence of children.

Visions, though, continue to be simply mystery, like another vision described in a poem I wrote on February 8, 1991:

> Darkened, that face of mine
> with the sleep of death,

forty years, frozen
tears did mummify

Words spoken from the mask,
tight lips did murmur,
Please release my soul
to fulfill my task.

An endless line, a circle made,
colored with rich bright rose;
across its girth a blackened blade
with a double edge.

It spoke to me of endless love
so rich, pure, and innocent,
it would carve in timeless space
the true strong sign of mother

Oh sister, if I could only
tell the beauty of that love
and our painful, giving task,
the true strong sign of mother.

Sever the golden chord you hold,
release your sons and daughters;
they are not for your comfort made
nor for your guilt to carry.

Give them instead the gift of rose
and with the blade across your breast,
endure your own magnificence;
live as their example.

There were fragments of that poem I could bring to rational, but in the light of my unfolding future, their significance fades. The most challenging phrase I had to live to was to endure my own magnificence. That vision and poem came when I was in a dark, dark place, before my upside-down, inside-out vision, and before my golden frog appeared. I have, though, come to accept and endure the gifts I have to offer, not only to my sisters. It is my hope that my revelations and my story will encourage a new vision and hope for our culture; encourage those seeking their own wholeness with a spirit that is free; encourage healing and freedom to those who are oppressed, abused, or in any way maltreated. As in Longfellow's poem, perhaps "a forlorn and shipwrecked brother, seeing shall take heart again." This brings me to another insightful moment in Poland, one that, in addition to owning my imperfections, has kept me modest.

My companion and I were walking together when I saw a man lying on the frozen sidewalk. He had brown hair and was wearing only a jacket. He was curled in a fetal position as though to keep warm; his brown tie-up shoes were placed neatly beside him, and I wondered why. I noticed that other people walking on the sidewalk simply curved around him. I must have stopped or hesitated, for I heard my companion's words, "Don't look!" as he pulled me away, but I could not help myself. It had been a moment, just a flash, and the millisecond that can bring numerous images before the words can even be formed: "How did my brother get to Poland?" And just as quickly, the thought, "There but by fate go I."[91]

I was somewhat aware of the belief that fit with an unconscious but deep conviction that people start from a place of innocence. It was at work in my practice as a therapist. It was drawn to my attention for the first time during my internship at a mental health center. I was working with a young man who related the horrendous abuse, neglect, and abandonment that marked his history. I protested his label of *schizophrenia* to my supervisor, for I saw no evidence in the therapy room of the symptoms recorded in his file. The supervisor replied, "You do not realize the effect you have on clients with the way you see them."

Like so many clients though, as I discovered, once labeled in the system, it is like a condemnation almost impossible to overcome. And when clients accept their label as a definition of who they are, it is over. They live out the script written for them.

That was the case of my client who died from an overdose of a prescribed psychotropic drug with horrific side effects on the body and the mind. His referral to me was for *maintenance* therapy, for no one expected a "schizophrenic" to improve. But over the years he did make improvements and became increasingly independent, requiring less and less medication. I listened to his symbolic language that gradually moved toward clarity. I supported him to work through some of the abuse he had encountered.

Throughout those years, though, I felt as if I was swimming against the tide, for he, like many irregulars, was the Identified Patient in his closed family and in the systems at large. My thought is that somewhere, at some point in his life, the ground fell out from under him and when under stress, his escape was to Spaceland, the world of symbols.

The biggest mistake I made was attempting to engage his family in the therapeutic process. After that, all the improvements gained began to decline. The news came the day his mother called to cancel his appointment, for he was in the hospital in a comatose state. He lived two more weeks and died on October 15th at 9 a.m., the fourth anniversary of our first appointment.

He was not in my conscious thoughts that morning. The most recent news was his physical condition was improving. I was engaged in doing my chores before meeting my afternoon appointments. I cleaned house, did the laundry, and then checked the clock. It was nine a.m. I went outside to work in my rose garden to spray, weed, and trim, then went back inside to check the time. It was nine a.m. After checking all the clocks in my house, both electric and battery, I called my daughter just to see if all was well. There had been many times when she needed me, and I would sense something amiss. I told her I was in some kind of time warp, and we laughed. A few hours later I was compelled to write the following poem:

Strange times,
moments filled
only with moments;
no shattering thoughts

of past or future
waiting, crushing, crunching
racing, chasing
that which
is not there.

Eternity begins
and ends here.

Clocks lose meaning
as they hesitate
and stop,
moving only
when attended
to meet demands
of time's
mechanical needs.

Three days later every rosebud in my garden blossomed, and I took 50 open roses to his burial.

I have no explanation for that beyond otherworldly experience except to accept it as another gift, a message to live in the present and realize that wonders abound when we are open enough to see them. Closely related is another of Jesus' maxims regarding the kingdom of heaven: "It will not come by watching for it. It will not be said, 'Look, here it is,' or 'Look, there it is.' Rather, the ... kingdom is spread out upon the earth and people do not see it."[92]

In many ways my life appears to be the same as it was before my transformation, for I still "must meet demands of time's mechanical needs," those responsibilities some refer to as the mundane. There is, however, no fog to prevent me from both the joys and sorrows of sensate existence. I am still impassioned when I see and hear of dishonesty, injustice, and oppression of any kind, especially the foul treatment of children, animals, and the helpless. I am still impassioned at human waste and the continued destruction of our earth.

In the keeping of my vows—facing my fears and breaking my silence—my life is oh, so different. The ghosts of my past have now been put to rest and the personal baggage I had accumulated through the years no longer weights me down. I am free from obsessions, for there is nothing to run from anymore. Free from the need for approval, free to make mistakes and correct them, the self-reflection and self-correction that is a part of living. I find joy in experiencing being completely human without illusions.

I have another painting hanging on my bedroom wall, one serving to remind me of my continuous human needs. It is a painting that also makes me smile, and it too, has many dimensions.

For over a year before leaving my special acre, I had worked, part-time, as a bookkeeper to make ends meet. For lunch, I would often treat myself to a Taco served at a nearby Quickie store. A young Asian woman, over time, came to know my order and never failed to look at me and smile. It was a smile I grew to cherish, for it exuded a gentle kindness. When there was an opportunity we would engage in friendly chat.

In October 1998, a month after my brother died, I had surgery to remove a melanoma on my neck. The next week I went to the canvas to paint, and the result was a gentle pensive face. It was a few days before I recognized the face I painted was the face of that kind woman. There had been no one to comfort me during my grief over Charlie's death or the anxiety I experienced during my illness.

That painting was symbolic of another deep longing of mine, to be treated with kindness, and I believe I recall every person throughout my life that did. In fact, after years of discovering the meaning of the symbols of my "box" dream, there was yet another dimension, something missing—the absence of kindness—food for the soul. I believe kindness is as rare to find as a truly honest man or woman. That gentle face has become an image reminding me to be kind and gentle to myself, which is very different from self- love—an omnipotent prideful place. It is not a place I have any desire to be. I see it all too often and arrogance, when unveiled for what it is, never paints a pretty picture.

Akin to the construct of self love in new age mythology is a *spiritual life*, which conflicts with the concept of continuum. All of life is permeated by energy/spirit. Those words are indicative of a

Sunday morning mentality, very different from living in the light. It is a way of life—the place where ethics and values meet—when one actually lives the principles I consider another continuum: honesty-kindness-justice-dignity and the wisdom of nature. A life that is honest dispels darkness, and kindness spreads the light that reveals injustices, gives value to another, and supports their dignity. I continue to make discoveries every day that evolve my thinking. And the most amazing part of such a life, its richness defies any measurement.

I am rich in being able to live in a new relationship with my daughter, separate, yet together, which is the purpose of individuation: to become whole in order to join with others without the heavy weight of dependency or need for control. Another aspect of that wholeness and living a balanced life is constancy. Moving from being a reactor to my environment to a sense of inner stability and self-control, the only kind of control one can strive for in acting on the world.

I am rich in being able to live in the moment with my daughter, grandchildren, sister, close friends, and acquaintances. I am rich in memory of a colorful life, including experiencing love with my companion; a confirmation of Alfred Lord Tennyson's poetic words, "Tis better to have loved and lost than never to have loved at all."

Yet I have a sense of privacy I had never known. No one knows what is in my mind and no one can enter my subjective world unless or until I share it. Apologizing when I make mistakes is a part of self-correction, but there is no need to confess my flaws to anyone but myself.

Another difficult lesson, and one I am still working on, is how to say *stop*—setting and maintaining personal boundaries—in a kindly and dignified way. While setting boundaries has many dimensions (enough to fill books and over two million sites on the Internet), it is a stance that says one has the right and the responsibility to protect oneself in relationships. Setting boundaries is easier as my values, honesty, kindness, justice, dignity and the wisdom of nature, become the criterion.

Setting boundaries also requires learning to say *stop* to myself on many levels in order not to be taken in by rhetoric or seduced by our materialistic Flatland. Even Jesus addressed that necessity with his words: "Be on guard against the world."[93] Thomas Jefferson was correct in saying, "The price of freedom is eternal vigilance."

Vigilance includes not allowing my mind to become closed but remaining open and flexible amidst ongoing change, for change is inevitable. Throughout all of my discoveries, though, one belief has remained constant. Evolving our cultures to open systems and nurturing the wholeness of our young is the "way" to achieve a safer and more egalitarian world where people are mindful of our life-sustaining earth and in caring for each other.

The *way* to nurture our future generations, though, is to be the person we want them to be. Regardless of what we try to teach them, they generally learn by example (modeling). Children are not only born innocent but open and vulnerable. They are not our possessions, yet it is our responsibility to protect them and to prepare them for life.

It is not my intent or calling to write a book on parenting. Any such book worth its salt, though, will address the space-time-matter-energy continuum. It will address the necessity of nurturing each of those dimensions to produce a new level of maturity—humans who can bring harmony to the world. It will address "reverence for life," including our life-supporting planet and our own.

Nurturing a rich imagination and self-expression through art, music, and language; nurturing fiscal and pragmatic responsibility through teaching them to take care of themselves. Nurturing them with healthy food, the stimulation of human touch, exercise, play, and relaxation; nurturing them with gentleness, for a child can never be loved too much, only overindulged. Supporting them in grievous times, acknowledging their feelings, and helping them to cope with whatever the luck of the draw has given them.

Another preparation for life is the teaching of boundaries—the cause-effect relationship—the consequences of mistakes and the need to self-correct. Nurturing the "most rare and excellent gift of modesty" and teaching the values of honesty-kindness-justice-dignity, together with the wisdom of nature.

Children are naturally curious and intuitive, a part of the instinct to not only survive, but to evolve. They can see, hear, and feel more than we give them credit for. They are adept in the language of symbols—the holistic functioning so often trained out of them. They are in a constant state of development unless they encounter the damage done by neglect, abuse, and closed system training.

Essentially we must nurture their wholeness then let them go to discover their own uniqueness and sense of purpose. John Raines[94] sums up that same idea succinctly:

> "In Daoism, before there is heaven and earth there is *Mother,* whose alias is Dao. She feeds all things and clothes all things but does not claim credit and does not master. Dao is ubiquitous and anonymous, conspicuous by its lack of ostentation. Here the story of life is one of generous giving, and then letting be and letting go. There is no demand for obedience, and there is no fall and no punishment. Most astonishing of all is a birthing that refuses to master, refuses to rule, and seeks instead only to nurture and mature."

The word, *Mother,* in the context of Dao (Tao), is neither masculine nor feminine but a state of wholeness. While nature does provide for all the creatures of the earth, including humans, the poor and hungry are evidence of man's "ultimate failure," to use Abbot's words, to divvy it up. As a mother I cannot begin to imagine what it would be like to have no food and to watch my child starve or die from a curable illness.

The souls relegated to darkness also lend validity to Abbot's words: "those who maintain superiority over multitude of their countrymen by their intellectual power are in conflict with nature and nature has condemned them to ultimate failure."

Raines' words, though, speak of the freedom of spirit, the way humans are meant to be. Free to evolve without fear and to make enlightened choices.

Even though *My Golden Frog* writing was predictive, I always had a choice to fulfill it or not, and at times I battled my dragons regarding that choice. Perhaps it was the corrective voice that provided the encouragement I needed: "Remember you are Bach, not Beethoven. Remember you are Bach, not Beethoven." Words so thick they could paint multiple images in my mind. My immediate response, and probably the most appropriate translation, was: *Listen, you are not deaf or mad, listen to the voice that lies deep within you.* They were words similar to and consistent with the line in one of my poems, "Put down the book, listen."

Though we may find validation and knowledge in books, it is our own inner voice that must take precedence. Perhaps one of the many meanings of Abbot's Spaceland poet's words: "To thine own self be true."

Longfellow's words were also helpful: "we can make our lives sublime." As I have found, the pursuit of happiness is an illusive thing, for the harder one chases, the more illusive it becomes. Sublimity is quite another story, and it comes softly with the joy wholeness brings. I hope to approach the end of my life as Leonardo da Vinci did: "As a well-spent day brings happy sleep, so life well used brings happy death."

I cannot imagine living in a world without music, poetry, or the beauty of nature. Part of that joy is expressed in living a life of gratitude for the gifts of my daily life, fully aware there are millions who are not so fortunate; gratitude for all those true heroes and creative people, including Edwin A. Abbot, who make "footprints on the sand of time."[96]

What I can imagine is a world where there is no fog, and human worth is not a matter of measurement. A world where all people have equal opportunity to mature and live full lives, those immeasurable higher dimensions to which we can aspire.

My vision of the "I" behind my insight now seems to have been a most wondrous gift on the 48th anniversary of my birth. *Colors of Dawn* is a brief piece of music providing shape and form for that vision, which foreshadowed a life that has turned inside-out and upside-down to what it was. What was dark and hidden has now become light. And to paraphrase a dictum by Jesus into the past tense: the two has become one; what was inside has become like the outside, and the upper like the lower; male and female has been made into a single one.[97]

Strangely enough, *Colors of Dawn* was written with the primary use of 5th intervals, which I later learned is anti-theoretical in the field of music, signifying the soul as the fifth dimension. The title could just as easily have been, *You Paint the Colors of my Soul.* "You" is in reference to Mother Nature and the Spirit that gives us life.

I was simply astonished at another song I wrote and titled *Relative Time*, for the music, if played at precisely 72 beats per minute, lasts exactly sixty seconds, a feat I could not have achieved with conscious intent. For a while I thought that *Relative Time* was an ode to Bach, but it had other dimensions. While I needed to write my own music and needed to become emotionally independent, it is a piece of music reminding me that it is not possible to be "fully human" alone. It also speaks of the harmony with which we can live. The moments of openness and sharing with people in my life, no matter how brief, are celebrated in the last phrase, "I remember you."

Yet another symbolic song, completed in 2002, has even more complex layers of meaning. It is titled, *Embrace the Mystery*. The complexity is in its constantly changing keys with a gradual movement toward fullness, the contents of my entire process in the form of a musical poem, and the ending was written long before I reached it. The song speaks of the mystery, the unknown and unknowable life-giving Spirit to which we are all connected in our souls.

The mystery of that spirit is evident in the song my granddaughter, Emily Grace, composed and sang to me with a gentle, lilting voice, just out of the blue, on October 14, 2009, when she was five years old. It was another moment of awe for me:

> Sometimes what is real is
> something you can't see.
> Invisible as the wind,
> hidden beneath the ground
> is the spring that feeds the creek.
> Sometimes your song
> can be a key to what is real
> but something you can't see.

I end my memoir with a sense of inner peace, for I have embraced the mystery.

## Appendix I

## My Rendition of Plato's *Allegory of The Cave*

Imagine people living in an underground cave with chains around their feet and head, fixed in a position so they cannot move. Only the cave wall in front of them is visible. Behind them, on a higher level is a fire, and between the fire and the prisoners are the rulers of the underground carrying all sorts of objects, intent on their stage activities, enacting the stories of the past. Those activities cast shadows on the wall in front of the prisoners. Some of those actors have voices, some of them are silent, and with the echo in the cave, the sounds seem to be coming from the shadows cast.

The chained inhabitants have spent their entire lives engaged in discussing those shadows, in debate, in agreement, entertaining their minds, ignorant of all except the stories evoked by the objects on the stage behind them. The actors, likewise, have no clue of anything except their own activities, unable to see the prisoners sitting in the darkness, even unaware of the shadows they cast.

Imagine for a moment that in a miraculous moment the chains around one's head and feet are suddenly gone and one can turn to see the fellow prisoners. Turning further one begins to see the actors and the objects on the stage, only dimly, though, for the light would be blinding to one so accustomed to darkness. Turning back to the only position one has ever known would be much more comfortable, for one could see the shadows much more distinctly. Surely it would take time to adjust one's vision, turning again and again, to examine that stage and the activities of the rulers, to realize where one has been all one's life, the illusions one has lived.

In that darkness, imagine what it would be like to see, at the top of a steep and rocky passage, another light shining through the opening of the cave. Even when one finally reaches that light, the brightness impinging on the senses would be overwhelming and the comfort of shadows and the light of the moon, more inviting. Adjusting to the world of light would also take time.

Imagine though, that one is plagued by the memory of the people still chained and what it was like to live in darkness. Re-entering that cave would seem another formidable task. It would take time to adjust to the darkness. At first one would stumble and fall, even endure the ridicule from the prisoners. In moments of doubt, one might even question if the world of light might have been the illusion except for the gifts one brings in one's soul.

## Appendix II

A narrative by Charles Andy Wall written seven years before his death on September 11, 1998, and included in his eulogy.

## Beauty is Here if We Look for It

I believe that 18 days from today I'll be released from jail. I believe I shouldn't be here for getting drunk and being an alcoholic, but I'm here.

I think that's my only negative statement of this letter. The purpose of my writing is to tell you of the faith I've gradually found in 41 years of living. Faith in God, I know is a matter of attitude in looking for the beauty to be seen in living day by day.

Believe me, it takes practicing a good attitude to see beauty sitting behind bars looking at pea-green walls.

I have pictures I've torn from magazines and tooth-pasted up in front of my steel leaf desk sticking out from the wall.

One is a pair of wood ducks whose beauty needs no explanation.

Another picture has a cat snuggling up to a fox that has been blinded by a car running over her.

The next has a big black circus bear sitting with a little boy looking over his shoulder at Thomas the Tank Engine and Friends magazine.

Another picture shows a fox trying with all his might to push an old fat goose in a baby carriage.

None of these pictures show fear or anger, only mutual respect for life and unselfish love. That love is the essence of my faith in God.

I was told as a child in Sunday school that God is love and he lives in the hearts of men everywhere. I believe that.

I have a picture of a little girl, barefooted, on crutches, leaning back on a stone wall that reminds me never to feel sorry for myself.

And when I look close at the pain and heartache in her little face, I can't help but shed a tear. For the gift God gave me to feel unselfish love and compassion, I am eternally grateful.

To the powerful faith I see in my picture of three polar bears swimming out to sea with not even an iceberg in sight, I will forever cling to. What I believe is that there's something good and beautiful in everything I see if I look for it. God made it that way.

P.S. I added another picture to my collection this morning. Set in royal hues of blue, it is of a guillotine, a social instrument of death and "justice."

I separated it by a good distance form the wood ducks and picture of the little fox trying to push the goose buggy.

Oh why, oh why must men conjure up ways to hurt each other instead of turning in to the beauty God made for us?

Bob Dylan sang the answer about the same time I graduated from high school. The answer my friend, is blowin' in the wind.

## Appendix III

# My eye opening experiences at a Drug Treatment Center in 1998

The Center was another closed system that impeded creativity or any dissenting voice. The cure rates no better than that obtained by chance.

And that chance had to do with "person" of the therapist to whom they were assigned. And that "person" had to do with the egos, or beliefs, comprised of the "socio-economic, political, cultural background, as well as the personality of those persons," to borrow Ellenberger's words. That chance also had to do with the "person" of the law enforcement or probation officer. The "person" of the judge, the "person" of the Department of Children and Family Services worker, the "person" of the doctor or psychiatrist.

It was so obvious when I walked into the corporate offices, then visited the treatment facilities, where the bulk of the money was spent. Grant money was the primary support of the Center, and one of the Grantor's conditions mandated that experienced, licensed therapists provide intervention. The licensed personnel who dealt with clients consisted of three, five-hour-per-week interviewers (one of them was me), and two other full-time licensed staff, one to sign off on treatments, the other to supervise. The actual intervention was performed by *cheap* labor, and more often than not, it was difficult to distinguish between the clients and the counselors.

The opportunity arose for me to work as a group counselor, a chance to put into practice the things I had learned in Poland and in my own process. I set to work to provide a model (syllabus) to include program requirements like HIV information, anger management, communication skills, parenting skills, social skills.

No model, though, can begin to describe the actual interchange as the group evolved by individuals working together. The use of color was introduced in the form of journals (that were respected as private), drawings, music, play, anything I could devise to allow and encourage free self-expression. There were few rules, and those rules had to do with boundaries, such as, "It is okay to have feelings of anger, it is not okay to act on them destructively," and, "No mind reading allowed." We worked on discovering personal themes and symbols in drawings, games, and interactions. I often reminded them, "If you don't talk it out, you will probably act it out."

I seldom addressed the symptom of addiction except for the required videos or unless the members brought it up, usually in confrontation of another group member. I honestly told them up front: "I offer no quick fix and will not compete with a bottle of Jack Daniels or any other drug."

I began each group by presenting drawings and information, teaching them everything I knew about consciousness and the existence of the unconscious, the need to express their thoughts, feelings, and explore their histories. I introduced relaxation techniques, encouraged exercise and healthy foods. One thing for certain: I did not ever tell them, or even suggest that they were powerless. Just as I suspected, there were few of them who were not interested in expressing their subjective worlds. I grew to love each of the members as individuals just as I had grown to love my clients in my practice. Relapses were rare. I became their advocate.

They were the homeless, the poor, of diverse races and cultural backgrounds. They were the progeny of the elite of society. What they all had in common was that each of them was the Identified Patient, not only of their family, but also of the system at large.

I often told them they were fortunate, for at least they *knew* they had problems and were learning to own them. They were well aware of the Flatland society in which they lived, well aware that the drug treatment program, like other healthcare programs, is big business. Its success is fixed with tainted research from questionnaires designed to shed good light, interested in the bottom line.

I learned quickly that it does not always pay to have a voice. I submitted letters challenging the decision to disband groups and refocus on education, a faster and cheaper method of intervention. In those letters I protested the proposed classroom programs and the fill-in-the-blank assessments that made objects of people. I confronted projections, gossip, and counselors who scapegoated clients. I refuted the demands for volume, not quality. Although I followed the rules of communication—no criticism without offering alternatives—none of my suggestions were even acknowledged. When I pointed out the success of members of my group, my supervisor responded, "You were working with a good group of people." And I was.

I was not fired. Part-time licensed positions were deleted after the grant money was acquired, and I was offered a full-time job with a schedule not requested of any other counselor: five evening shifts each week (at half the hourly pay). The catch was I had to abide by their new rules of education. I stayed long enough to bring closure with group members then resigned.

I wrote a final letter confronting their fraudulent practices, and during the last week I was at the Center, I saw teams of people, including the administrator, going through all the client files with pens in hand. I strongly suspect they were "doctoring" them.

# Endnotes

[1] Yalom, Irving D., *Love's Executioner & Other Tales of Psychotherapy*, Perennial Classics, 1989

Nietzsche's words were quoted in the preface. In later editions that quote had been deleted.

[2] Storr, Anthony, *The Essential Jung*, Princeton University Press 1983.

[3] *A Psalm of Life* by Henry Wadsworth Longfellow:

> Tell me not in mournful numbers,
> Life is but an empty dream!
> For the soul is dead that slumbers,
> And things are not what they seem.
>
> Life is real! Life is earnest!
> And the grave is not its goal;
> Dust thou are, to dust thou returnest,
> Was not spoken of the soul.
>
> Not enjoyment, and not sorrow,
> Is our destined end or way;
> But to act, that each tomorrow
> Find us farther than today.
>
> Art is long, and Time is fleeting,
> And our hearts, though stout and brave,
> Still, like muffled drums, are beating
> Funeral marches to the grave.
>
> In the world's broad field of battle,
> In the bivouac of Life,
> Be not like dumb, driven cattle!
> Be a hero in the strife!

Trust no Future, howe'er pleasant!
Let the dead Past bury its dead!
Act, – act in the living Present!
Heart within, and God o'erhead!

Lives of great men all remind us
We can make our lives sublime,
And, departing, leave behind us
Footprints on the sand of time;

Footprints, that perhaps another,
Sailing o'er life's solemn main,
A forlorn and shipwrecked brother,
Seeing, shall take heart again.

Let us then be up and doing,
With a heart for any fate;
Still achieving, still pursuing,
Learn to labor and to wait.

[4] Campbell, Joseph, *The Hero With a Thousand Faces*, New World Library, 2008.
[5] Jung, Carl G., *Man and His Symbols*, Dell Publishing, 1968.
[6] *Plato, Five Great Dialogues*, Walter J. Black, Inc., 1969.
[7] Abbott, Edwin A., *Flatland, a Romance of Many Dimensions*, Dive, 1952.
[8] Thomas F. Banchoff, *Beyond the Third Dimension, Geometry, Computer Graphics, and Higher Dimensions*, Scientific American Library, 1990.
[9] Shlain, Leonard, *Art & Physics, Parallel Visions in Space, Time & Light*, William Morrow and Company, Inc., 1991.
[10] Edwards, Betty. *Drawing on the Right Side of the Brain*, Jeremy P. Tarcher/Putnam, 1989.
An expanded and updated edition was published in 1999: *The New Drawing on the Right Side of the Brain*, a product of Betty Edwards' evolving thought.
[11] Britt, Laurence W., *Free Inquiry magazine*, Volume 23, # 2.
[12] Betty Edwards suggests that one of the functions of the left brain (linear) is symbol production like letters formed to make a word or a number to

indicate quantity. I use the word, *symbol*, differently, as a function of the right or spatial hemisphere production, usually an image or images that can have many layers of meaning—multidimensional in its character—and can only be interpreted by the producer.

[13] December 24, 2008. Psychiatrist, and former Harvard professor, Dr. Diane Hennacy Powell in an online Time interview by M. J. Stephey.

[14] Guiley, Rosemary Ellen, *Harper's Encyclopedia of Mystical and Paranormal Experience*, New York: Harper Collins, 1991.

[15] Smith, Adrian, *The Harvard Crimson*, Super symmetry and Parallel Dimensions, Harvard Physicist (Lisa) Randall among world's leading string theorists. January 6, 2006.

[16] Whitehead, A. N., *Nature and Life*, Cambridge, 1934.

[17] Koestler, Arthur, *The Roots of Coincidence*, Vintage Books Edition, 1973.

[18] Jung, Carl G., *Man and His Symbols*, Dell Publishing, 1968.

[19] Chesler, Phyllis, *Women and Madness*, Harcourt Brace Jovanovich, 1989.

[20] Theodore Roosevelt.

[21] This image of chains around one's head is taken from Plato's famous metaphor of *The Cave* which I elaborate in Appendix I.

[22] Gross, Jan Tomasz, *Neighbors: The Destruction of the Jewish Community in Jedwabne, Poland*, Princeton University Press, 2001.

[23] Ellenberger, Henri F., *The Discovery of the Unconscious: The History and Evolution of Dynamic Psychiatry*, Basic Books, 1970.

The title is a misnomer, for artists and poets have known of the unconscious for thousands of years. There is a story in the Bible of Joseph, famous for his coat of many colors, who was freed from imprisonment through his comprehension of prophetic dreams. It is a story that fascinated me, even as a child, just how much, I was unaware, for it entered my symbolic world. My mother nurtured a plant called "Joseph's Coat." It was a plant from which I took cuttings and transplanted to my gardens wherever I moved 1961-2000. I did not recognize its meaning for me until 1993. The plant was symbolic of my hope to be freed from my imprisonment through my own dreams.

[24] Jung, Carl G., *Man and His Symbols*, Dell Publishing, 1968.

[25] Pinker, Steven, *How the Mind Works*, W.W. Norton and Co., 1997.

[26] Stout, Martha, *The Paranoia Switch*, Sarah Crichton Books, 2007.

[27] Gould, Stephen Jay, *The Mismeasure of Man*, W.W. Norton & Co., 1981.

[28] Farb, Peter, *Humankind; What we know about ourselves, Where we came from and where we are headed. Why we behave the way we do.* Bantam Books, 1978.

[29] Hobson, Art, *Physics and Human Affairs*, John Wiley & Sons, Inc., 1982.

[30] Banchoff, Thomas F., *Beyond the Third Dimension, Geometry, Computer Graphics, and Higher Dimensions*, Scientific American Library, 1990.

[31] Gleick, James, *Chaos*, Penguin Books, 1987.

[32] Hawking, Stephen, *A Brief History of Time, From the Big Bang to Black Holes*, Bantam Books, 1990.

[33] Schwenk, Theodor, *Sensitive Chaos, The Creation of Flowing Forms in Water and Air*, Rudolf Steiner Press, London, 1965.

[34] Shlain, Leonard, *Art & Physics, Parallel Visions in Space, Time & Light*, William Morrow and Company, Inc., 1991.

[35] Capra, Fritjof, *The Science of Leonardo, Inside the Mind of the Great Genius of the Renaissance*, Doubleday, 2007.

[36] My mind's eye sees the constant, almost imperceptible, expansion and contraction of that body of matter as it breaths, emitting and receiving energy, a living entity. My mind's eye sees the earth's conception in a burst of energy and the teeming activity of particles taking form through the forces of the universe as though responding to some vast genetic program for roundness. I see the earth maturing to give birth to life in its myriad specie, continuing evolution as it rumbles and quakes, constantly changing its surface.

I think how similar I am to the earth, my body primarily fluid, plasma flowing as rivers dividing into tiny rivulets, nourishing me in an orderly rhythm; how my own form reflects the flowing forms of all living things. I think of how I carry in my body the genetic material reflecting billions of years of evolution; how I, too, am subject to the forces of the universe, my body affected by the tides and the phases of the moon. I, too, matured to give birth to life, and my surface changes with each passing year.

I contemplate the miracle of being human, the capacity to hold the complex image of the earth in my mind, the capacity to reflect time, the order of the universe and to reflect space where past, present, and future can exist simultaneously.

[37]Agee, James, *A Death in the Family*, Bamtam Books, 1982.

> James Agee began, *A Death in the Family*, his autobiographical novel in 1944, but it was not quite complete when he died in 1955. It was edited and released posthumously in 1957. It reveals the long-term effects of trauma in a child's life.

[38]Stewart, Ian and Joines, Vann, in their book, *TA Today*, Lifespace Publishing, 1987.

> Stewart and Joines contend there is no such thing as a victim and further assert that clients made decisions in infancy and, as children, planned their whole lives. Such a ludicrous statement, for the brain of an infant is in a constant state of development, both biologically and psychologically, and that development depends, primarily, on its environment.

[39]Morrison, Toni, *The Nobel Lecture in Literature*, 1993, Alfred A. Knopf, 1994.

[40]*Plato, Five Great Dialogues*, Walter J. Black, Inc., 1969.

Plato's Seventh Book of *The Republic*.

[41]Dee L. R. Graham, *Loving to Survive, Sexual Terror, Men's Violence and Women's Lives*, New York University Press, 1994.

> Graham's work has been hampered, though, by extremism. She goes too far when she asserts that even a heterosexual identity is a result of conditioning. Advocating lesbian relationships as an alternative to dependence on men is ludicrous, as though true homosexuality is a choice. Graham attempts to categorize and know the minds of other women based on *research*. I suspect her subjective confession is at the root of "being trapped by her rhetoric and belief in her own objectivity."

[42]I propose it is the failure of bonding or an early disruption of that initial bond, that is at the root of the more serious psychological disabilities, especially those who do not develop a sense of inner stability, e.g. the so-called borderline personality disorder.

[43]Many years later, my aunt's daughter told me she moved because she could not tolerate seeing what was happening to us children, particularly me; whether true or not, it was a validation of my childhood nightmares.

[44]Charlie hanged himself in the Sumter County jail.

[45]Morrison, Toni, *The Nobel Lecture in Literature*, 1993, Alfred A. Knopf, 1994.

[46]Bookkeeper, legal secretary with a specialty in real estate; three years as a bookkeeper for the Episcopal Diocese of Central Florida, 10 years employment with an Interior Designer as a bookkeeper/business manager.

[47] *The Diagnostic and Statistical Manual of Mental Disorders*, published by the American Psychiatric Association (in conjunction with drug companies), has been revised four times, and is currently in process of being revised again.

[48] One only need refer to the *Citizens Commission on Human Rights, Investigating and Exposing Psychiatric Human Abuse* website to get a taste of what is truly going on.

[49] Dunigan, Sinead, *Devastation via Inoculation*, dissidantvoice.org 2006.

Drug companies deny any connection between injected toxins with autism, and it remains a controversial issue. Perhaps it is like their decades of denial of the harm that prescribed hormones have done to women. Despite drug manufacturers' denial, Dunigal presents compelling evidence linking the upsurge in autism with the mercury contained in children's inoculations.

[50] Numerous writings regarding childhood trauma and "poisonous pedagogy" published between 1978 and 2009. Alice Miller died on April 14, 2010, at age 87.

[51] A Florida Mental Health Act, 1971, that can be used for involuntary commitment to a mental health facility. Its initial purpose was to incarcerate the elderly by family members in order to gain control over their funds. It has also served those whose desire was to be *rid* of bothersome irregulars or spouses. The editor of my biography had been incarcerated in a mental institution for a decade by such a husband who had influence in the *right* places.

[52] Graham, Dee L. R., *Loving to Survive, Sexual Terror, Men's Violence and Women's Lives*, New York University Press, 1994.

[53] *Silence of the Lambs*, a 1991 film based on the novel of the same name by Thomas Harris.

[54] Capra, Fritjof, *The Science of Leonardo, Inside the Mind of the Great Genius of the Renaissance*, Doubleday, 2007.

[55] Research on neural plasticity has focused on pathology, the damage imposed by trauma, abuse, abandonment.... That work needs to be extended to a more hopeful level, the capacity of the human to heal.

[56] "The unexamined life is not worth living" was Socrates' statement at his trial for heresy. An enigmatic sage, philosopher, and teacher, best known through Plato's writings, encouraged his students, through his method

of questioning, to challenge the accepted beliefs of the time and think for oneself. It was a stance which he was unwilling to give up and it cost him his life.

[57] *The I Ching or Book of Changes*, The Richard Wilhelm Translation rendered into English by Cary F. Baynes, Princeton University Press, 1990.

[58] "Pushing upward has supreme success. One must see the great man [woman]. Fear not. Departure toward the south brings good fortune." Pushing upward uses the metaphor of a tree pushing upward out of the soil, a pushing upward that comes only with unrelenting effort. Departure toward the south—hard work—brings good fortune. The greatest challenge was seeing my own inner worth, necessary for realizing my potential.

What was most amazing, though, that which made a believer of me, the chances of throwing six coins, six times, and receiving the same pattern, (heads and tails) over and over, was far, far beyond mere coincidence.

[59] I learned the basic rudiments of music from an elderly lady in Linden. I had to quit taking lessons; the only reasons I recall is that her charge increased from 50 cents to a dollar, and I did not practice enough. My sister, however, went on to another teacher in Webster. I used her books to teach myself.

[60] I was fortunate I had resolved my familial past. That resolution helped me to deal with and care for my mother with kindness during a two-year period of requiring round-the-clock care, and her subsequent death on January 29, 2003.

[61] McHugh, Paul, Book Review, The Wall Street Journal, June 29, 2008, reviewing Richard J. McNally's Book, *Remembering Trauma*, Belknap/Harvard, 2003.

[62] Frank, Jerome D. and Frank, Julia B., *Persuasion and Healing: A Comparative study of Psychotherapy*, John Hopkins University Press, 1991.

[63] Huang, Alfred, *The Complete I Ching*, Inner Traditions International, 1998.

[64] Gould, Stephen Jay, *The Mismeasure of Man*, W.W. Norton & Company, 1981.

[65] There are over 3 million references on the Internet regarding differences between male and female brains.

[66] Voneida, Theodore J., *Roger Wolcott Sperry*, The National Academies Press. www.nap,.edu/html/biomems/rsperry.html

[67] Yalom, Irvin D., *Existential Psychotherapy*, Basic Books, 1980.

In his book, prominent psychiatrist, Irving Yalom, discounts and ridicules clients for their use of the *I Ching*. In the same vein, Yalom states that in his psychotherapy he does not pay attention to client's dreams unless they contain "a loud voice." He contends, throughout his subjective confession, that the basic existential problem is fear of death, then goes on to discount any humane sense of purpose other than his own, but fails to reveal what that might be.

I am in agreement with some of Yalom's thoughts, especially that it is dehumanizing to label people, that human behavior is not genetic but oftentimes "begins so early it feels genetic." He, too, is aware of the disturbing number of therapists who are as damaged as their clients. I am fully in agreement with Yalom's belief in the necessity of exploring one's history as evidenced in *Love's Executioner*, cited at Endnote one. It becomes obvious, though, that Yalom has little comprehension of symbols and relies heavily on language, at which he is especially adept. I can also appreciate his honesty and self-reflection evidenced by his awareness of his own prejudices.

[68] Meyer, Marvin, *The Gnostic Gospels of Jesus*, Harper Collins Publishers, 2005.

[69] Pagels, Elaine H. *Beyond Belief: The Secret Gospel of Thomas*, Random House, 2003.

[70] Pagels, Elaine and King, Karen L., *Reading Judas, The Gospel of Judas and the Shaping of Christianity*, The Penguin Group, 2007.

[71] Over a million and a half websites on *lucid dreaming*.

[72] Morrison, Toni, *The Nobel Lecture in Literature*, 1993, Alfred A. Knopf, 1994.

[73] Wilber, Ken, *A Brief History of Everything*, Shambala Publications, Inc., 1996.

[74] The Cult of Ken Wilber's Integral Institute—over 1,000,000 sites on the Internet.

[75] Burns, Rhonda, The Secret, Atria Books, 2006.

[76] Compartmentalize—To separate into distinct parts, categories, or compartments: "You learn ... even the ability to compartmentalize ethics." Ellen Goodman.

[77] Miller, Alice, *Banished Knowledge*, Doubleday, 1990.

[78] It was a reminder of an incident with my mother when I asked her about my childhood physical punishments. Her response was, "Why Frances, I never laid a hand on you!"

[79] Matthew 25: 40.

[80] Matthew 7: 7.

[81] Maguire, Daniel C. and Shaikh, Sa' diyya, *Violence Against Women in Contemporary World Religion*, The Pilgrim Press, 2007.

[82] Farb, Peter, *Humankind; What we know about ourselves, Where we came from and where we are headed. Why we behave the way we do.* Bantam Books, 1978.

[83] Kidder, Tracy, *Mountains Beyond Mountains*, Random House, Inc. 2004.

It is a book that describes a man who is inspiring, and daring. It is a book that will open one's eyes to truths that are truly disturbing.

[84] Farmer, Paul, *Pathologies of Power, Health, Human Rights, and the New War on the Poor*, University of California Press, 2005.

[85] Lawrence, Robert S., Johns Hopkins University, in praising *Pathologies of Power*.

[86] Brabezon, James, *Albert Schweitzer: A Biography*, G. P. Putnam's Sons, 1975.

[87] Edwards, Betty. *The New Drawing on the Right Side of the Brain*, Jeremy P. Tarcher/Putnam, 1999.

[88] Leonard Shlain refers to this shift to left brain dominance in his book *The Alphabet Versus the Goddess: The Conflict between Word and Image*. (Viking-Penguin, 1998).

[89] Elbow, Peter, *Writing With Power, Techniques for Mastering the Writing Process*, Oxford University Press, 1981.

[90] Edwards, Betty. *The New Drawing on the Right Side of the Brain*, Jeremy P. Tarcher/Putnam, 1999.

As Betty Edwards suggests, our educational systems need to be more "whole-brained" in orientation, and school curriculum needs to give equal weight to the arts and creativity.

[91] My original words were, "There but by grace go I," one of those mindless statements suggesting that "God" favors one person over another.

[92] The Gospel of Thomas 113: 2 (I edited out father's).

[93] The Gospel of Thomas 21: 6.

[94] Maguire, Daniel C. and Shaikh, Sa' diyya, *Violence Against Women in Contemporary World Religion*, The Pilgrim Press, 2007.

[95] William Shakespeare.

[96] From Henry Wadsworth Longfellow's Psalm of Life.

[97] The Gospel of Thomas 22: 4.